MASCULINITY IN
FOUR VICTORIAN EPICS

*In memoriam, J. W. Machann (1917–1965)
and Sophie Machann (1918–2009).*

Masculinity in Four Victorian Epics
A Darwinist Reading

CLINTON MACHANN
Texas A&M University, USA

ASHGATE

© Clinton Machann 2010

All rights reserved. No part of this publication may be reproduced, stored in a retrieval system or transmitted in any form or by any means, electronic, mechanical, photocopying, recording or otherwise without the prior permission of the publisher.

Clinton Machann has asserted his right under the Copyright, Designs and Patents Act, 1988, to be identified as the author of this work.

Published by
Ashgate Publishing Limited
Wey Court East
Union Road
Farnham
Surrey, GU9 7PT
England

Ashgate Publishing Company
Suite 420
101 Cherry Street
Burlington
VT 05401-4405
USA

www.ashgate.com

British Library Cataloguing in Publication Data
Machann, Clinton.
    Masculinity in four Victorian epics: a Darwinist reading.
    1. Tennyson, Alfred Tennyson, Baron, 1809–1892. Idylls of the king. 2. Browning, Elizabeth Barrett, 1806–1861. Aurora Leigh. 3. Browning, Robert, 1812–1889. Ring and the book. 4. Clough, Arthur Hugh, 1819–1861 – Criticism and interpretation. 5. Epic poetry, English – History and criticism. 6. English poetry – 19th century – History and criticism. 7. Masculinity in literature. 8. Darwin, Charles, 1809–1882 – Influence. 9. Social Darwinism in literature. 10. Literature and science – Great Britain – History – 19th century.
    I. Title
    821'.03209353'09034-dc22

Library of Congress Cataloging-in-Publication Data
Machann, Clinton.
    Masculinity in four Victorian epics: a Darwinist reading / by Clinton Machann.
       p. cm.
    Includes bibliographical references and index.
    ISBN 978-0-7546-6687-5 (alk. paper)
    1. English poetry—19th century—History and criticism. 2. Masculinity in literature. 3. Narrative poetry, English—History and criticism. 4. Darwin, Charles, 1809–1882—Influence. 5. Tennyson, Alfred Tennyson, Baron, 1809–1892. Idylls of the King. 6. Browning, Elizabeth Barrett, 1806–1861. Aurora Leigh. 7. Clough, Arthur Hugh, 1819–1861. Amours de voyage. 8. Browning, Robert, 1812–1889. Ring and the book. 9. Darwin, Charles, 1809–1882—Influence. 10. Social Darwinism in literature. I. Title.

PR508.M35M33 2010
821'.809353—dc22

2009036617

ISBN: 9780754666875 (hbk)
ISBN: 9780754699897 (ebk)

Printed and bound in Great Britain by
TJ International Ltd, Padstow, Cornwall

Contents

Acknowledgments		*vi*
1	Introduction	1
2	Tennyson's Arthur and Manly Codes of Behavior	31
3	Barrett Browning's Construction of Masculinity in *Aurora Leigh*	57
4	Clough's Ambivalent Victorian Manhood	83
5	Browning's Chivalrous Christianity	109
6	Conclusion	141
Bibliography		*149*
Index		*161*

Acknowledgments

My fascination with Victorian literature has been central to my entire career as a teacher and scholar. Over the years my publications have focused to a large extent on the poetry and critical prose of Matthew Arnold, and since 1994 I have contributed each year an essay on Arnold to the annual "Year's Work" series in the journal *Victorian Poetry*. However, it was writing about Tennyson that inspired me to undertake the study that has led to this book, and my 2000 essay "Tennyson's King Arthur and the Violence of Manliness" (*VP* 38:200–26) was an early version of Chapter 2. I want to thank *VP* editor John Lamb for his permission to use the essay in this way. Gender studies, specifically the study of masculinity, has also been a longstanding interest of mine, and among my essays and reviews published in the *Journal of Men's Studies*, "The Male Villain as Domestic Tyrant in *Daniel Deronda*: Victorian Masculinities and the Cultural Context of George Eliot's Novel" in 2005 (*JMS* 13.3: 327–46) contains some references to Browning's *The Ring and the Book* that I have incorporated into Chapter 5.

As for my use of ideas and interpretive strategies associated with literary Darwinism, I have become intensely interested in that critical approach in recent years, and I am grateful to Joseph Carroll for his pioneering work and for the opportunity to discuss with him important ideas concerning literature, evolutionary psychology, cognitive studies, and related issues. I am pleased to say that Joe visited our campus here at Texas A&M to deliver a lecture and visit my literature classes in 2003, and he was followed by two other prominent scholars in the field who did the same thing: Brian Boyd came in 2006, and Nancy Easterlin in 2008. I was delighted to meet and visit with them, as was my friend and colleague Brett Cooke, a specialist in Russian literature, who himself has significant publications in biopoetics, closely associated with literary Darwinism. Brett and I together taught a graduate seminar entitled "Darwinian Approaches to Literary Studies" in the spring of 2007, and I appreciate and enjoy our continuing collaboration on plans for related activities in the future.

Looking back much further, personal experience – spending my youth in an agricultural, "farming and ranching" environment – helped me to understand, intellectually and emotionally, from an early age that indeed human beings are part of the natural world, that biology transcends human culture and language, long before I became acquainted with scientific theories about biological evolution. This book is dedicated to the memory of my mother and father.

Finally, the sympathetic patience and intellectual support and advice of my wife Ginny have been invaluable during the long period in which I worked on this book.

Chapter 1
Introduction

This is a study of the representations of masculinity – the trait of behaving in ways thought to be typical of or appropriate for males – in four important Victorian long poems: Alfred Tennyson's *Idylls of the King* (various publishing dates, 1842–91), Elizabeth Barrett Browning's *Aurora Leigh* (1857), Arthur Hugh Clough's *Amours de Voyage* (1858), and Robert Browning's *The Ring and the Book* (1868–69). "Ideal manhood closed in real man," alluding to King Arthur, is a phrase Tennyson added to the epilogue of his poem in 1891, as he made his final corrections.[1] Questions of gender are of the greatest concern to Tennyson, as they are to Barrett Browning, Clough, and Browning, and the topic of masculinity is central to our understanding of Victorian literature: its major themes, its idealism and social criticism, its perplexities and uncertainties.

In this chapter I review the importance of the long poem as a literary genre in mid-to-late nineteenth-century British literature, with an emphasis on the classic poems represented here. Then I discuss my critical and theoretical assumptions about masculinity in the context of literary gender studies and, at a more basic level, about the function of literature. In particular, I outline my use of concepts and methods associated with "literary Darwinism," a relatively new critical approach to literature that in my view offers a fruitful way to explore issues related to the traditional concept of "human nature" that have been largely ignored since the advent of "postmodernism" and the prevailing assumption that literature – like all cultural formations – is "socially constructed." I retain the term *construction*, especially when referring to gender, because it has been ubiquitous in gender studies, with the understanding that social or cultural constructions are not independent of human nature.

In recent years, open-ended discussions about the continuing relevance of Victorian studies (especially studies of Victorian poetry and within that subdivision especially studies of the "long poem") have been prominent in the field, questioning the direction such studies should take in the future. As postmodernist literary theories lose some of their momentum, scholars explore a variety of historicist approaches while taking care to maintain a general allegiance to the political and feminist values that have been dominant in literary studies for the past few decades and avoid a return to traditional, "humanist" approaches or what are now often seen to be misguided assumptions about the coherence or unity of "Victorian" culture symbolized by Britain's longest reigning monarch.[2] In advocating a

---

[1] According to his son, Hallam Tennyson, as discussed in Chapter 2.
[2] See, for example, the variety of approaches discussed in the two special issues of *Victorian Poetry* (somewhat whimsically) entitled "Whither Victorian Poetry?" 41.4 (2003) and 42.1 (2004). These issues were edited by Linda K. Hughes, who provides an editorial introduction in each case.

serious consideration of literary Darwinism as we revise our approach to Victorian poetry, I am asking that readers and critics be willing to recognize and take into account a body of scientific literature by anthropologists, evolutionary scientists, cognitive scientists, and other researchers about the "human condition" that has been accumulating over the years. Like most others who have recognized this important contribution, I assume that individual human volition, social culture, and a "universal" human nature work together in complex ways.

The term *literary Darwinism* might be misleading to some who are unfamiliar with this "rapidly evolving" critical tradition because Darwin was not only an important scientist whose landmark contributions to theory and research will always be associated with evolution but also an author whose literary voice is a vital ingredient in British Victorian culture. For example, in *Darwin's Plots: Evolutionary Narrative in Darwin, George Eliot and Nineteenth-Century Fiction* (1983), Gillian Beer studies Darwin's narrative "plots" in *The Origin of Species* and other books and shows how George Eliot, Thomas Hardy, and other novelists were influenced by Darwinian versions of "life struggles" in plotting their novels. Of course, Darwin exerted an enormous influence of this kind, and his "cultural relevance" to Victorian literature in general, including works by the poets whom I discuss in this book, is important. In a broad sense, Darwin shares the Victorian cultural scene with Tennyson, Barrett Browning, Clough, and Browning. Furthermore, Darwin's own literary style is engaging and continues to charm readers today; however, as noted above, the concept of literary Darwinism is based on a constantly expanding body of research – in the tradition of Darwin – concerning how our knowledge about the adapted human mind is related to the study of literature.

Literary Darwinists do not believe in "genetic determinism," nor do they assume that the evolutionary process has ended, but they do offer compelling reasons for the continuing interest today in, for example, Homer and the ancient Hebrew scriptures, not to mention British Victorian literature, in various national and ethnic contexts and after many generations of cultural change. They help to explain the universality of themes and motifs in contemporary literatures across the world, the remarkable connections we find among various literatures, and the fundamental value of comparative literary studies and, simultaneously, interdisciplinary literary studies that combine the insights gained by those literary studies with the findings of biological and social scientists. They help us understand why literature of the distant or recent past retains a powerful potential for engaging the interest of students – and the general adult reading public – in ways that may seem surprising today, when academic discourse about literature has become increasingly isolated from anyone outside the specialized disciplines of the university.[3] After outlining

---

[3] In a 2007 article that focuses on Lionel Trilling's perceptive – and still relevant – 1955 essay about Matthew Arnold's poem "The Scholar-Gipsy," David Rampton notes that critics who published studies of Victorian poetry during the period 1950–75 still "had a very significant audience for their work, one that extended far beyond the confines of the discipline" but that now "literary criticism has almost completely lost whatever popularity it once had outside the academy" (13).

my reasons for believing that both the genre of the Victorian long poem and representations of masculinity in Victorian literature are deserving of new critical attention at this time, I discuss my critical assumptions in greater detail.

Victorian Long Poems

The four poems announced in my title remain generally familiar to students of Victorian literature – after all, one was based on a "national myth" and occupied the efforts of the most celebrated poet of the age throughout his career, one was an immensely popular poem in its day and has now re-emerged as a feminist classic, one expressed characteristic doubts and confusions of the age in a particularly effective way, and one is the acknowledged masterwork of the poet whose experimental style most influenced poets (and novelists) in the twentieth century. However, it has become increasingly difficult for readers and critics to understand and appreciate the prominence of the long poem among the Victorians. In fact, it might be generalized that Victorian poetry as a whole has had a smaller audience of readers and critics in recent decades,[4] although selected short poems by poets persist as assigned readings from British literature anthologies used in university classrooms. (Of course, in some cases, anthologized poems are excerpts from the long poems.) I hope to encourage fresh readings of these poems that fully recognize their historicity as expressions of "Victorian" ideas and attitudes while appreciating their continuing relevance as explorations of deep and timeless human issues. To do this implies a renewed appreciation for traditional, "humanist" scholarship, and in the course of my discussion below I make some references to this work, but always within a context that rejects misguided assumptions about the fundamental incompatibility between literary and scientific scholarship and instead assumes a consilience in our search for knowledge about human culture and the natural world.[5]

In the late twentieth century some important scholarship in the field of Victorian poetry focused on the formal or generic experiments of poets faced with dilemmas in extending both Romantic (lyric) and classical (lyric, narrative, dramatic) modes of poetic expression within a dynamic aesthetic and critical environment

---

[4] In a rough survey of MLA Bibliographies beginning in the early 1960s (referring to the standard MLA headings), I found, for example, that the total number of entries for Tennyson scholarship climbed from 384 during the decade 1963–72 to a peak of 526 in 1973–82 and that there was a decline to 423 in 1983–92 and 296 in 1993–2002. The rise and fall of (Robert) Browning studies was even more dramatic, from 425 in 1963–72 to a peak of 653 in 1973–82, and then a rapid decline to 427 in 1983–92 and 244 in 1993–2002. This trend has continued.

[5] "Consilience," referring to a synthesis of knowledge from different specialized fields and efforts to unite the sciences with the humanities, is a key term in biologist E. O. Wilson's *Consilience: The Unity of Knowledge* (1998).

increasingly given to social or cultural critique.[6] In the case of the long poem with
overt or implied pretensions to the genre of epic, the pressures on the Victorian
poet were particularly severe. Matthew Arnold is a well-known example of a
conflicted Victorian poet who spontaneously composed poetry in the mode of the
Romantic lyricist but (in the opinion of most of his major critics) rejected his own
best efforts to compose long poems[7] while persistently attempting to recapture a
classical poetry of the "grand style" in dramatic and epic narrative verse. Arnold
himself accepted the idea that poetic form changes in response to cultural change,
and although in his early lectures as Oxford Poetry Professor he applied himself
to the perplexing question of how best to translate Homer for a Victorian British
audience, he went on to develop his critical prose based on the idea that criticism
must help prepare for an age in which great poetry – like the epic and dramatic
works of ancient Greece – could be written once again.

Beyond his classicist idealism, Arnold, like most Victorian poets, yearned to
write substantial, that is, *long*, poems, and many of his contemporaries with less
purist views of classical models adapted them freely in combination with various
historical, Romantic, and innovative modes. The final version of Tennyson's
Idylls of the King consists of 10,289 lines of blank verse organized into 12 books.
Barrett Browning's *Aurora Leigh* is of similar length – 10,938 lines of blank verse
in 9 books – and Browning's *The Ring and the Book* is about twice as long: 21,116
lines of blank verse in 12 books. Clough's *Amours de Voyage*, 1,243 lines of
unrhymed hexameters, is by far the shortest of the four poems. As we will see later
in extended discussions of these poems, all four authors associated their works
with "epic" generic traditions, with Clough's poem as a kind of "mock-epic."
But even Browning's massive poem was only about half the length of
P. J. Bailey's *Festus*. This version of the *Faust* story, featuring scenes with God

---

[6] Among the most influential studies of this kind have been Isobel Armstrong, *Victorian Poetry: Poetry, Politics, Poetics* (1993); W. David Shaw, *The Lucid Veil: Poetic Truth in the Victorian Age* (1987); and E. Warwick Slinn, *The Discourse of Self in Victorian Poetry* (1991).

[7] Many critics today consider Arnold's 1852 "lyric–drama" *Empedocles on Etna* (1,121 lines of blank verse) to be one of his most successful and important works: he saw the plight of the ancient philosopher as analogous to that of the Victorian intellectual, with his scepticism, acute self-consciousness, sense of isolation and loneliness, nostalgia for a lost world. But Arnold soon became dissatisfied with this poem because it did not "charm" or "delight" the reader and because the suffering it portrays finds no vent in action. He excluded the poem from his 1853 collection of poems and sought to justify this exclusion in his famous critical "Preface" to that volume. Other (moderately) long poems published by Arnold include *Sohrab and Rustum* (892 lines of blank verse, the 1853 poem of classical simplicity and action that in a sense supplants *Empedocles*) and *Tristram and Iseult* (789 lines of various meters). Arnold's was the first modern treatment of the Tristram myth in English, published along with *Empedocles* in 1852, 30 years before Algernon Charles Swinburne's much longer *Tristram of Lyoness* (4,468 lines of heroic couplets). In some ways, however, the most intriguing long poem planned by Arnold was one about the Roman poet and philosopher Lucretius that he never completed. (See Machann, *Matthew Arnold*, 24, 27, 40–41, 43, 71–2.)

and angels in Heaven, was originally published in 1839 and then revised – and expanded – in subsequent editions: the 11th "Jubilee" edition of 1889 consisted of about 40,000 lines of blank verse. In his day, Bailey, whose reputation rested almost entirely on *Festus*, was associated with the "Spasmodic" school of poetry, faulted by some critics for a characteristically morbid, psychological intensity,[8] and of course he is almost entirely unread today; nonetheless, he had a large contemporary audience. An even longer poem, apparently the longest published poem of the century, was by a poet who remains canonical: William Morris's *The Earthly Paradise: a Poem*, in four volumes, 1868–70. Morris, still known today for his designs and paintings as well as his poetry, was strongly influenced by Geoffrey Chaucer, and the 25 individual narrative poems that make up the whole are various in metrical form (much like the *Canterbury Tales*), adding up to a grand total of 42,681 lines.[9]

A closer look at one of the Spasmodic poems, however, will help us to understand the remarkable popularity but also volatility of the genre of the Victorian long poem, and it will serve to contextualize important aspects of the major works under consideration in this study. William Edmondstoune Aytoun coined the critical term *Spasmodic* in an 1854 mock review in *Blackwood's Magazine* of his own imaginary poem, a parody entitled *Firmilian*. He subsequently completed and published this parody, which he called a "Spasmodic tragedy."[10] Along with Bailey, Alexander Smith and Sydney Dobell were poets associated with this "school" whose work came to be widely ridiculed as a result of Aytoun's effective critique, and Smith's career in particular, based on the outstanding initial success of his long poem "A Life-Drama" (3,086 lines), first published along with shorter works in his *Poems* (1853), suffered a dramatic reversal.

The construction and critical fate of "A Life-Drama" have important implications for the discussions of the long poems by Tennyson, Barrett Browning, Clough, and Browning, in Chapters 2–5. For that reason, I offer here a short discussion of Smith's remarkable career.

The son of working-class parents who lived in Kilmarnock and later moved to nearby Glasgow, he followed his father into the textile trade and became, like him, a designer of calico printing and sewed muslins. Although Smith had little formal education, his mother taught him Gaelic songs and Ossianic legends, and his youthful love of poetry led him to read the English Romantics and then Tennyson with enthusiasm. In 1850, at the age of 20, he began to publish his own poems in the Glasgow *Evening Citizen* and developed friendships with other young men

---

[8] A recent revival of critical interest in the Spasmodics is illustrated by a special issue of *Victorian Poetry* (Winter 2004), including essays by Herbert F. Tucker, Jason R. Rudy, Kirstie Blair, Linda K. Hughes, Antony H. Harrison, Charles LaPorte, Emma Mason, and Florence S. Boos.

[9] Morris's gigantic collection of tales based on Greek and Norse mythology is structured by a narrative device similar to that used by Chaucer in *Canterbury Tales* and does not possess a unified plot like that of *Festus*.

[10] See Mark A. Weinstein, *William Edmondstoune Aytoun and the Spasmodic Controversy* (1968).

in Glasgow with literary interests, including Thomas Brisbane, who would later write his biography.[11] When Smith submitted a collection of his poems to the influential critic George Gilfillan in Dundee, Gilfillan praised them and helped to publish some of them in the *Eclectic Review*, but he advised Smith to write a longer poem that would make a stronger public impression. Smith responded by constructing a narrative that allowed him to incorporate some of the poems he had already written. The result was "A Life-Drama," which made him famous – at least for a few years.

The protagonist in this poem is Walter, a poet with an aristocratic background, whose ambition is to write a poem that will make him famous: "Poesy! Poesy! I'd give to thee, / As passionately, my rich-laden years / ... / As Hero gave her trembling sighs to find / Delicious death on wet Leander's lip" (6) and "O Fame! Fame! Fame! Next grandest word to God! / I seek the look of Fame!" (9). Not surprisingly, a love interest is added to Walter's literary ambitions. While he is sleeping in the woods on his estate, a mysterious Lady comes by and is attracted both to the young man and to the book of poetry that lies by his side. Their mutual love of poetry is the catalyst to their brief romance, which is doomed from the beginning by the Lady's commitment – vaguely suggested – to marry a man she does not love, and apparently she dies of unhappiness after she marries him, as she expects she will. Two years later, Walter, still brokenhearted, is in London, where he tells his sad story to his friend Edward. Edward invites him to his manor, where he introduces him to Violet, the beautiful young woman who becomes Walter's second love. Like the first Lady, Violet is enchanted by Walter's recitation of poetry as well as poetic ambitions, but apparently Walter takes advantage of her during their courtship and violates her sexually – at least that is what is implied in his conversation with a prostitute he encounters on a bridge at midnight. After an absence of three years, a repentant Walter returns to his home, resolving to begin his life anew. A conversation between his friends Edward and Charles reveals that the "great poem" that Walter had planned from the beginning has been published and has met with success, but he remains unhappy until he is rejoined with Violet at the end of the poem. In the final scene the two have been together watching a sunset and then the emerging moon and stars – formerly associated with Walter's poetic engagement with "the infinite" – but as they turn to enter Walter's house, he remarks that "A star's a cold thing to a human heart, / And love is better than their radiance" (160).

As suggested in this brief synopsis, the plot of the poem is not well developed or even fully coherent. The 13 "scenes" that make up the poem contain a few dramatic exchanges between Walter and the other characters, but neither his relationships with the two women in his life nor with those with his friends are explored in any depth. All too clearly, Smith constructed the flimsy plot as a device to string together a series of lyric moments in which the poet Walter responds emotionally to the natural world and expresses his romantic feelings for the two women. Furthermore, there are narrative poems and "songs" recited by Walter – and by

---

[11] *The Early Years of Alexander Smith, Poet and Essayist* (1869).

the women – within the larger narrative that have a kind of allegorical function. In Scene IV Walter has been reading to his own Lady a poem about a Page in love with a Lady: "I drop the mask; / I am the sun-tanned Page; the Lady thou!" (65). In Scene VIII, Walter tells the tale of a suffering young poet who yearns for fame but is frustrated in his ambition and loses confidence in himself, and when Violet asks "Did you know well that youth of whom you spoke?" he answers, "Know him! O, yes, I knew him as myself –" (113). Surely very few readers were surprised by Walter's autobiographical revelation in either case, but such devices tend to serve as digressive interludes and distract the reader from the dramatic situations. Smith, of course, incorporated previously composed poems in this way.

Nevertheless, early readers of "A Life-Drama" responded favorably to the sensuous imagery in the Romantic tradition incorporated into this poem, and indeed some individual passages are striking. In Scene V, Walter watches the sunset as he walks down a rural lane:

> The flying sun goes down the burning west,
> Vast night comes noiseless up the eastern slope,
> And so the eternal change goes round the world.
> Unrest! Unrest! The passion-panting sea
> Watches the unveiled beauty of the stars
> Like a great hungry soul. (70–71)

Here he is discussing his favorite topic of poetry with his friend Edward:

> To set this Age to music, – the great work
> Before the Poet now. I do believe
> When it is fully sung, – its great complaint,
> Its hope, its yearning, told to earth and heaven, –
> Our troubled age shall pass, as doth a day
> That leaves the west all crimson with the promise
> Of the diviner morrow, which even then
> Is hurrying up the world's great side with light.
> Father! If I should live to see that morn,
> Let me go upward, like a lark, to sing
> One song in the dawning! (85–6)

Among the admirers of Smith's poetry were George Meredith and Herbert Spencer, who was "strongly inclined to rank him as the greatest poet since Shakespeare" (Duncan, *Life and Letters*, 67), and Smith's *Poems* had already reached a fourth edition by 1856. When Clough reviewed the 1852 volume of his friend Matthew Arnold's poems, he compared Arnold's poems unfavorably with those of Smith. At this point, Clough admired Smith's work and was impressed by the achievements of this Scotsman despite his working-class background.[12] Arnold was angry with

---

[12] In fact, Clough's admiration for Smith's poetry was short-lived, fading soon after the 1852 review, but Charles LaPorte points out what he thinks are affinities between Clough's own poetry and that of Smith and other Spasmodics in his essay "Spasmodic Poetics and Clough's Apostasies."

his old friend, especially since he did not respect Smith's work. In the famous "Preface" to his 1853 volume of poems, in which he explains why he had omitted his own long poem "Empedocles on Etna" from that volume, Arnold makes references to the unfortunate "dialogue of the mind with itself" in "Empedocles" – a feature he found to be characteristic of the age and that he was determined to avoid in his own work.[13] As shown in their correspondence, Arnold and Clough had longstanding disagreements about the proper subject matter of poetry, and I return to this issue in Chapter 4. Smith's extreme subjectivity was a strongly negative example for Arnold; however, it was Aytoun who made the devastating attack on Smith and the other "Spasmodics," and soon the majority of critics tended to take a negative view of Smith's (sometimes morbid) psychological intensity, passionate subjectivity and egoism, and emphasis on the physical body. Public opinion about Smith's poetry shifted quickly.

In an 1853 letter, Barrett Browning wrote that although Smith "has noble stuff in him," he "has more imagery than verity, more colour than form" (*Letters* II, 120). Nevertheless, her own *Aurora Leigh* is sometimes associated with the Spasmodic school, and certain features of that poem can be compared with "A Life-Drama" in ways that suggest an interesting influence, so I will have more say about the "Spasmodic" Smith in Chapter 3, as well. In a larger sense, the example of Smith illustrates the emphasis on the long poem as a genre – and the special status of the poet in mid-Victorian culture. Smith, who acknowledged the stylistic excesses of his first volume, spent the rest of his career trying to recapture his youthful fame, but he had only moderate success as an essayist and poet. He turned to a historical subject in his final volume of poetry, *Edwin of Deira* (1861), a poem of 3,275 lines in relatively straightforward, unadorned blank verse that treats the story of the Northumbrian king who, according to the Venerable Bede, ruled from 617 to 633 and founded the city of Edinburgh. Clearly, Smith had set aside his Spasmodic style, but critics pointed out his apparent imitation of Tennyson, who published his first version of the *Idylls* in 1859.

Tennyson had of course established himself as a poet with the publication of *In Memoriam A. H. H.* (1850), the long collection of elegiac lyrics occasioned by the sudden death of his close friend Arthur Henry Hallam in 1833. Smith admired Tennyson, who had successfully molded his series of personal, short lyric poems into one of the major long poems of the age (and built the reputation that secured his position as Poet Laureate). When Smith's initial success with his own generic leap from short lyrics to long narrative poem turned suddenly into failure, he attempted to follow Tennyson's lead in writing the story of a historical or mythic king, but the new direction of his work was widely seen as unoriginal and did little to restore his popularity.

Overall, however, the long poem was a strong, vital, and popular literary genre throughout the Victorian period. In addition to *Idylls* and *In Memoriam A. H. H.* (2,900 lines), Tennyson published *Maud* (1,325 lines), *Enoch Arden* (911 lines), and *The Princess* (3,309 lines). Besides *Aurora Leigh*, Barrett Browning

---

[13] See Machann, *Matthew Arnold*, 35–9.

produced *The Battle of Marathon* (1,462 lines – written at the age of 13), *An Essay on Mind* (1,262 lines), and *Casa Guidi Windows* (2,002 lines). Clough's poetic reputation was based largely on *The Bothie of Tober-na-Vuolich* (1,732 lines) and *Dipsychus* (2,020 lines, unfinished), in addition to *Amours de Voyage*. Most prolifically, Browning, after *Paracelsus* (4,151 lines) and the notorious *Sordello* (5,982 lines),[14] went on to publish *Christmas Eve and Easter Day* (2,399 lines), *Balaustion's Adventure* (2,705 lines), *Prince Hohenstiel-Schwangau* (2,155 lines), *Fifine at the Fair* (2,463 lines), *Red Cotton Night-Cap Country* (4,247 lines), *The Inn Album* (3,079 lines), *Aristophanes' Apology* (5,711 lines) and *Parleyings with Certain People* (3,495 lines), in addition to *The Ring and the Book*. And other titles by other poets found large readerships. Hundreds of thousands of copies of *Festus* were sold as it ran through 11 editions, for example, and *The Earthly Paradise* was a popular success as well. Victorian readers did not simply tolerate long poems: they had a special taste – and respect – for them,[15] and poets strove for the prestige and moral authority traditionally associated with the genre. Of course, the productions by Tennyson, Barrett Browning, Clough, and Browning focused on here not only retain their significance in historical terms for their status among contemporary nineteenth-century audiences but continue to fascinate modern readers because of their technical experimentation and their negotiations with cultural and ideological issues, although they would not serve to rejuvenate the long narrative poem as a popular genre in the long run. Historical studies such as Matthew Reynolds's *The Realms of Verse 1830–1870: English Poetry in a Time of Nation Building* (2001) remind us that the poems featured in this book were very much involved in a complex rethinking of important issues involving cultural and political unity and nationalism.[16] Nevertheless, today it is easy to forget that these works were written, published, and read in a time when the long poem was widely

---

[14] Browning's *Sordello*, the product of nearly 10 years' work, is a historical poem set in twelfth- and thirteenth-century Italy. It had the reputation of being one of the most obscure and confusing poems of the early Victorian period, and Browning's reputation suffered from this negative reception for many years, although twentieth-century critics were more sympathetic to Browning's struggles with poetic meaning in the poem.

[15] Adam Roberts's *Romantic and Victorian Long Poems: A Guide* (1999) is a particularly useful handbook that incorporates descriptions and summaries of British long poems (approximately 1,000 lines or longer) published during the nineteenth century. In his introduction, he discusses the popularity of the long poem during this era, emphasizing the significance of epic traditions. He notes that "Victorian literature is some of the most prolix in the canon. It is a shame, indeed, that 'prolix' carries such negative connotations, because it is part of the Victorian literary project that it was able to dilate upon its subject" (9).

[16] I do not mean to suggest that the long poem has disappeared as a literary genre today. On the contrary, although the genre has lost its dominant status, a few long poems continue to find enthusiastic readers. For example, in a recent article John P. Farrell compares Arnold's own *The Scholar-Gipsy* as a "celebration of the continuous life of poetry" with the modern long poem *The Continuous Life* (1990) by Mark Strand. A special issue of *PMLA* "On Poetry" (120.1, January 2005) calls attention to uncertainties in the current status and significance of poetry in the widest sense.

assumed to be the most prestigious literary genre, but we must remind ourselves of that fact when we study Victorian literature in its historical context.

The Ring and the Book has long been considered the most interesting of these poems as a cultural experiment. The genre of dramatic monologue, pioneered by Browning, is widely assumed to be the most significant contribution of Victorian poetry to modern literature. Here a series of monologues, already a hybrid of lyric and drama, is based on the plot of a lurid crime story drawn from original documents dating from seventeenth-century Rome and extended to epic proportions in a way that implies the unity of experience and discourse for critics; it inspired further experimentation with interior monologues and stream-of-consciousness techniques by twentieth-century novelists. More recently, Barrett Browning's *Aurora Leigh*, which she considered a modern epic and a novel in verse and which enjoyed the reputation of being a major work before suffering from neglect by post-Victorian critics, has also been discussed as a major cultural and literary experiment, with emphasis on gender – Barrett Browning's construction of the self-conscious autobiographical voice of the woman poet, and modernity – her insistence on contemporary settings and the representation of contemporary social and political issues.

Barrett Browning deliberately placed her work in opposition to Tennyson's Arthurian epic project:

> I do distrust the poet who discerns
> No character of glory in his times,
> And trundles back his soul five hundred years,
> Past moat and drawbridge, into a castle-court,
> To sing – oh, not of lizard or of toad
> Alive i' the ditch there, – 'twere excusable,
> But of some black chief, half knight, half sheep-lifter,
> Some beauteous dame, half chattel and half queen,
> As dead must be, for the greater part,
> The poems made on their chivalric bones;
> And that's no wonder: death inherits death. (V, 188–98)

and

> King Arthur's self
> Was commonplace to Lady Guenever;
> And Camelot to minstrels seemed as flat,
> As Regent Street to poets. (V, 209–12)

However, from the time that Tennyson, at the age of 24, began to plan his series of "idylls" based on the Arthurian legends, he himself had been keenly aware of the risks he was taking and uncomfortable with the "epic" generic label. In his narrative frame "The Epic," consisting of a sort of prologue and epilogue to the first version of "Morte d'Arthur" (later entitled "The Passing of Arthur" as the final book of the *Idylls*), he has his epic poet–protagonist Everard Hall attempt to burn all 12 books of his Arthurian epic:

"Nay, nay," said Hall,
"Why take the style of those heroic times?
For nature brings not back the mastodon,
Nor we those times; and why should any man
Remodel models? These twelve books of mine
Were faint Homeric echoes, nothing-worth,
Mere chaff and draff, much better burnt." (34–40)

Hall's friend fortuitously rescues one book, the "Morte," from the hearth, and after the poet reads it aloud to a group of friends, the unnamed speaker remarks, "Perhaps some modern touches here and there / Redeemed it from the charge of nothingness" (328–9). Tennyson's defensiveness is evident and can be compared to Barrett Browning's concerns about her ambiguous identity as a woman poet and her poem as a curious modern epic or a novel in verse and Browning's anxieties about his reputation as an odd and obscure writer as well as the radically experimental form of his work. For his part, Clough, in the voice of his protagonist in *Amours de Voyage*, openly expresses doubts and uncertainties that he knew would be associated with his own life, and his innovative adaptation of the hexameter gave his verse a distinctive quality indeed. Of course, I return to these issues in later chapters.

Although Tennyson did not complete his epic poem until about five years after the publication of Browning's *The Ring and the Book*, I think it is appropriate to discuss Tennyson first because he had begun to publish parts of the work, already conceptualized as a 12-book epic, 15 years prior to the publication of Barrett Browning's *Aurora Leigh*. In contrast to *Amours de Voyage*, *The Ring and the Book,* and *Aurora Leigh*, *Idylls of the King*, although clearly "experimental" in its linking together generically uncertain "idylls" (quite unlike the idylls of Theocritus) in order to create a modern, classically epic-length version of an ancient Celtic myth best known in its Medieval prose version, was from the beginning seen as essentially archaic by both its admirers and detractors. Modern critics, while not ignoring Tennyson's vision of the rise and fall of civilizations, generally do not categorize it as a cultural experiment to be compared with *The Ring and the Book* and *Aurora Leigh*. I will, however, suggest that *Idylls* is in some ways speculative and experimental in ways that have not been fully appreciated.

The relationship between *The Ring and the Book* and *Aurora Leigh* is also complex and interesting, and some of the links between them are discussed in Chapters 3 and 4. Although Browning dedicated his epic poem to the memory of his wife Elizabeth, who had been dead for about seven years by the time he published it, there is evidence that Barrett Browning did not much like her husband's preliminary plans for the poem.[17] Furthermore, it has been argued that one of Browning's fundamental aims in his poem was to metaphorically "kill" his dead wife again and appropriate her poetic vision by identifying her with the

---

[17] See William Irvine and Park Honan, 409.

saintlike, victimized heroine Pompilia.[18] Clearly this claim is related to issues of gender roles represented in the two poems.

Beyond specific issues of intertextuality among the poems is a shared thematic and formal concern with tensions between historicity and an ahistorical or universal vision. In *Idylls*, Tennyson is ambivalent about associations between historical time and great mythic cycles, between the current need for British leadership within its world empire and the heroic ideal of the once and future king. In *Aurora Leigh* Barrett Browning indeed focuses on the contemporary world, but her historically situated poet–heroine in the tradition of Romantic metaphysics places her trust in a transcendent God and experiences an apocalyptic vision of the future. In *Amours de Voyage* Clough emphasizes ties between the contemporary Roman setting and the ancient city of history and myth. In *The Ring and the Book* Browning offers a narrative based on archival documents from a particular historical moment and, as Carol Christ puts it, "represents man's consciousness as determined by history" but at the same time "implies a point outside of history from which that consciousness can be judged and the structure of history comprehended."[19] In the discussion below, I show how my own critical approach deals with the tension between historically (or culturally) based literary interpretations and those that imply "universal" human experiences. In this sense my work is in the tradition of recent studies that have analyzed the complex relationship between the historical present and ancient historical or mythic traditions in long poems by Victorians attempting to adapt an epic vision: Colin Graham's *Ideologies of Epic: Nation, Empire and Victorian Epic Poetry* (1998), Simon Dentith's *Epic and Empire in Nineteenth-Century Britain* (2006), and Herbert F. Tucker's *Epic: Britain's Heroic Muse 1790–1910* (2008).

Tucker's massive work is especially helpful in demonstrating the pervasive presence of epic ideas in major and minor poems of the nineteenth century, a factor given little attention by mainstream literary critics. He treats three of the major poems discussed here as major Victorian epics, discussing Barrett Browning's success in channeling the "spasmodic impulse" into "the most important epic poem" of the decade 1850–60 (340), Browning's decision "to make a major Victorian epic out of the trial documents he found one day bound together for on a bookstall in Florence" (437), and Tennyson's reluctance to adopt the term *epic* but then "making an epic virtue of the pluralist necessity to which the title *Idylls of the King* committed him" (455). However, "generic scruple" prevents him from

---

[18] An extreme form of this argument is made by Nina Auerbach in her article "Robert Browning's Last Word" (1984). According to Auerbach, "Having survived a poet who had made epic claims for herself, Robert Browning perpetuated her voice by turning it into his own: he 'married' Elizabeth Barrett one more time when he appropriated her after her death, weaving her declarations into the corrosive fabric of his dramatic monologues" (173).

[19] Carol Christ, *Victorian and Modern Poetics*, 114. Christ's chapter "Myth, History, and the Structure of the Long Poem" (101–41) is particularly insightful in its discussion of how Victorian poets such as Arnold, Tennyson, and Browning strive to reconcile historical and ahistorical elements in their works and how they anticipate the modernist experiments of Ezra Pound, T. S. Eliot, and William Butler Yeats.

discussing Clough's, "non-heroic," "anything-but-epic" *Amours de Voyage* (338, 390). There is no doubt that Clough systematically rejects an epic vision, but in doing so he alludes to epic traditions in virtually every line of his odd, English hexameters, and, as already indicated, I discuss Clough's poem as a "mock-epic."

Graham, Dentith, and Tucker all offer insights into the interactions among genre, ideology, and historical consciousness. However, my own critical approach is somewhat different and, I hope, will supplement these previous studies in important ways. Because I focus on issues related to masculinity, it is useful to review briefly the way in which the study of gender as a specialized area of critical discourse was introduced and integrated into the field of Victorian literary and cultural scholarship.

Victorian Studies and Concepts of Masculinity

In his widely read and influential book *The Victorian Frame of Mind, 1830–1870* (1957), Walter Houghton suggests that his study of the "general character" or temperament of the age has a special significance for his readers "because to look into the Victorian mind is to see some primary sources of the modern mind" (xiv). Houghton expressed his opposition to prevailing negative and patronizing attitudes toward the Victorians, but even earlier works like Jerome H. Buckley's *The Victorian Temper, A Study in Literary Culture* (1951) and John Holloway's *The Victorian Sage, Studies in Argument* (1953) had begun to create an academic vogue for Victorian studies at least partially based on the premise that the ideas, values, and life philosophies of early-to-mid Victorian intellectuals and artists anticipated fundamental intellectual and aesthetic problems of the mid-twentieth century. A decade later, the popularity of works like George Levine and William Madden's edition of essays *The Art of Victorian Prose* (1968) demonstrated the continuing force of this idea, at least in English literature graduate programs in American universities.

By the early 1970s, the now well-established tradition of the special relevance of Victorian literature was incorporating new voices that over the next two decades would profoundly transform the field of Victorian studies. Houghton, like many of the other scholars taking what now is seen as a traditionalist, humanist approach to Victorian literature, had focused on ideological and moral dilemmas associated with the sudden changes brought on by industrialization, urbanization, religious ferment, the growth of science and technology, and so on. Now, by focusing on gender issues, feminist scholars were beginning to interrogate the ways in which Victorian literature and culture had been studied. Elaine Showalter, Sandra Gilbert, Susan Gubar, Martha Vicinus, Mary Poovey, Margaret Homans, Judith Newton, and Cora Kaplan are only a few of the important writers who helped to make feminism dominant in Victorian studies by the 1990s. Although feminist scholars had been associated with a variety of innovative theoretical methodologies, from Marxism to Lacanian psychoanalysis and the poststructuralism of Derrida and Foucault, it was the ideology of feminism and the methodological emphasis on gender study that drove this transformation. Like Houghton, Victorianists were still fascinated by the prospect of finding clues to contemporary culture in literary productions of the Victorian era, but now this project was conceived in terms of understanding how gender is socially constructed and how gender issues are central to all literary and cultural discourse. Antony Harrison and Beverly Taylor's representative 1992 edition of essays entitled *Gender and Discourse in Victorian Literature and Art* was grounded in the belief that, by the nineties, feminist critical inquiry, although far from monolithic, offered a coherent, highly-developed, and, in fact, dominant approach to Victorian literature in contemporary academic discourse. Since that time, mainstream feminist assumptions have been widely accepted as foundational and rarely challenged.

Central to the feminist program is the proposition that the supposed universal human nature postulated by Victorian poets, sages, and novelists (and by Houghton and other traditional Victorianists for that matter) is a false universal, one closely associated with the experience of white European men (but not women) of a certain class and with a patriarchal social order that embodies instabilities, contradictions, and hypocrisies of various kinds. It is generalized that "with regard to institutional modes of discourse and the power structures of Victorian society, women for the most part constituted a suppressed and marginalized underclass whose victimization was a political fact of life resisted in diverse ways or complexly exploited by artists sensitive to the relations among power, authority, and gender" (Harrison and Taylor, xii).

However, even if one accepted the idea that feminine experience was systematically ignored or suppressed within the old paradigm of supposed universal human experience, it was by no means clear that gendered masculine experience had been adequately or authentically explored within that paradigm. It could be argued that issues of masculinity had been suppressed as fully as feminist ones in traditional humanist discourse. Men's studies focusing on homosexuality (because they too could be seen as running counter to heterosexual male traditions) were the first to be legitimatized by the feminist enterprise, but then a few analyses of "mainstream" masculinities appeared. A prominent example is Herbert Sussman's study of Victorian masculinities (1995). Sussman's work is to some extent based on Eve Sedgwick's scheme of homosocial desire (1985) and, like much feminist, poststructuralist criticism, ultimately derives from the history of sexuality as formulated by Michel Foucault, but Sussman rejects a monolithic model of heterosexual male experience and attempts to analyze the conflicts and anxieties he finds in the masculinity of traditionalist writers like Thomas Carlyle. Sussman's use of the plural *masculinities* "stresses the multiple possibilities of such social formations, the variability in the gendering of the biological male, and the range of such constructions over time" (8). James Eli Adams, in another study of Victorian male authors from the mid-1990s, similarly analyzes "competing constructions of normative masculinity within a single historical moment" (11).[20]

---

[20] In *Dandies and Desert Saints: Styles of Victorian Masculinities*, Adams, referring to Sussman's focus on Victorian male brotherhoods that emphasize manliness as self-control, argues that his own approach "adopts a more external perspective on the figures it analyzes, focusing on masculine identity as a social logic, a strategy of self-presentation" (12).

Traditional British Marxist approaches to history and culture have been particularly influential in studies of Victorian gender, and Thaïs E. Morgan, for example, in a 2000 overview of the "poetry of Victorian masculinities," specifically makes use of Raymond Williams's "model for analyzing the competing discourses that constitute culture" (204).

Houghton and other "humanist" Victorianists conventionally divided the larger "age," beginning with the Reform Bill of 1832 and ending with the death of Queen Victoria, into three periods: the early period or Time of Troubles (approximately 1832–48), characterized by social ferment and the Condition of England Question; the Mid-Victorian period (approximately 1848–70), a time of economic prosperity or complacency and religious controversy; and the late period (approximately 1870–1901), marked by aestheticism and decadence or the decay of Victorian values. Versions of a tripartite scheme adapted by literary men's studies as a branch of feminist gender studies have included the following formulations: (1) early Victorian attempts to define a masculine ideal through ascetic discipline (Carlyle's monasticism, Evangelical and Tractarian versions of a life devoted to higher values and usefulness); (2) more aggressive mid-Victorian models of masculinity associated with British imperialism, notably Kingsley's "muscular Christianity," which celebrated animal spirits and sexual energy as well as robust physical strength and enjoyment of life while retaining a need for self-discipline and self-denial; and, finally, (3) late Victorian eclecticism, which accommodated Pater's "aesthetic historicism" and finally saw the emergence of a previously forbidden homoeroticism. (It is assumed of all three periods of course that even normative masculinities are "multiple, complex, and unstable constructions.")[21]

Later, another important aspect of Victorian masculinity was studied in some depth. John Tosh's 1999 book *A Man's Place: Masculinity and the Middle-Class Home in Victorian England* attempts to show why domesticity was closely linked to masculinity in Victorian England. Tosh observes that "never before or since has domesticity been held to be so central to masculinity," and he examines family records and didactic texts in an effort to reconstruct "how men of the Victorian middle class experienced the demands of an exacting domestic code, and how they negotiated its contradictions" (1). It is remarkable that stereotypes related to "the doctrine of separate spheres" for men and women in Victorian middle-class culture had for so long obscured the central importance of home and family in the lives of Victorian men.

In spite of the attempts of a few scholars like Tosh to study Victorian "masculinities" in a broad cultural context, the emphasis for the most part has been on literary convention and the gender consciousness of the author, "the development of 'masculine poetics'" (Sussman, 14). And this development is seen to be rather rapid, one "historical moment" quickly displacing another. Going beyond the three stages outlined above, Morgan generalizes, "Representations of masculinity... shifted several times in Victorian England" (203). This hyperhistoricist categorization

---

[21] See Adams, 3.

of literary and cultural convention sometimes obscures writers' attempts to deal with more fundamental aspects of human experience. Today's literary theorists tend to be suspicious of Victorians' attempts to explore "human nature"; after all, that concept itself is often assumed to be a linguistic and cultural construction, even if one acknowledges that attempts to accommodate human experience and human identity to dynamic and troubling new concepts of the natural world was one of the dominant concerns of Victorian writers and intellectuals. In the case of gender, especially, there is a general reluctance to associate either the masculine or the feminine with biological sex differences. Sussman is expressing a widely understood academic anxiety when he reassures his readers that:

> I use the term "male" only in the biological sense and the term "maleness" for fantasies about the essential nature of the "male," for that which the Victorians thought of as innate in men. I reserve the terms "masculinity" and "manliness" for those multifarious social constructions of the male current within society. Thus, using "masculine bonding" rather than "male bonding" or "masculine poetic" rather than "male poetic" suggests the constructed rather than the essentialist, the diverse rather than the monolithic nature of these formations. Such a distinction is especially important for the Victorians for whom the hegemonic bourgeois view defined "manliness" as the control and discipline of an essential "maleness" fantasized as a potent yet dangerous energy. (12–13)

I want to make it clear that although I take a somewhat revisionist stance in my own analysis of "Victorian masculinities" in this study, I begin by recognizing the genuine value of studies such as those by Sussman and others whose critical assumptions are similar to his. I, too, use the terms *masculine*, *masculinity*, and *masculinities* to refer to socially constructed, historically specific definitions of gender. I also find much that is useful in the observations of those critics – and here Adams is especially prominent – who stress the performative dimension of masculinity: "normative masculinity is ... asserted as an unending performance" (11).

However, this does not mean that I am a "pure" constructionist because I assume that a prior biological reality exists and that gender differences ultimately derive from an interaction of innate qualities and cultural concepts. For example, Tennyson's anxieties about an innate potential for violence in men, discussed at length in the next chapter, become cultural constructs when incorporated into the *Idylls*, and they can be compared to those expressed in certain passages from his *In Memoriam*, among the best known quotations from Victorian literature: "Are God and Nature then at strife, / That Nature lends such evil dreams?" (55:5–6) and "Nature red in tooth and claw" (56:15). But culture can be seen as a product as well as a shaper of human desires, and the rapidly developing fields of Darwinian or adaptationist psychology and cognitive science have been enhancing our understanding of what it is still appropriate to call "human nature." This is not a matter of interpreting all cultural phenomena as the direct effect of genetics or of reducing either group culture or individual human experience to biology but rather of linking biology, psychology, the social sciences, history, and culture in an effort to understand their complex interactions. In the case of male violence, recent studies of prestate societies show that 10 to 60 percent of men die at the hands of other men. The psychologist Steven Pinker notes that violent "cultures of honor" emerge in "just about any corner of the world that is beyond the reach of the law" and cites the opinions of historians who "argue that people acquiesced to centralized authorities during the Middle Ages ... to relieve themselves of the burden of having to retaliate against those who would harm them and their kin." He suggests that "the growth of these authorities may explain the *hundredfold* decline in homicide rates in European societies since the Middle Ages.[22] And yet, with an apparently expanding civilization and moral consciousness, the threat of anarchy and violence on a vast scale remains. Tennyson's critique of Victorian idealism can be profitably studied within this paradigm. Furthermore, the Arthurian Order of the Round Table as conceptualized by Tennyson is a fraternal and ideological coalition designed to impose authority and control resources. Social mores imposed by the Order attempt to suppress the expression of core social behaviors – not only regulating male-on-male violence through an elaborate code of rules that undercuts such violence as a direct mode of competition for the sexual favors of women (even to the point of promoting chastity as an ideal to be associated with valor) but by suppressing polygamy and otherwise controlling relations between the sexes.[23] However, the controlled violence of the tournament (along with various male initiation rituals, one variety of masculine *performance*) is still the primary means of achieving masculine status within the coalition.

Tennyson's treatment of male violence can be compared to Clough's seemingly casual references to both contemporary and ancient warfare and Browning's emphasis on male brutality and criminality in *The Ring and the Book*. In a larger sense, issues related to male violence are central to conflicted Victorian attitudes toward the valorization of the "primitive" and the "barbarous" in the ancient epics, as emphasized by Dentith. This kind of ambivalence of the Victorians toward the epics of Homer, for example, is easily understandable when viewed in the context of modern studies that analyze the pervasive violence of "Homer's world."

---

[22] See Steven Pinker, *The Blank Slate: The Modern Denial of Human Nature*, 70. Pinker summarizes the findings of M. Daly and M. Wilson, *Homicide* (Hawthorne, NY: Aldine de Gruyter, 1988); L. H. Keeley, *War Before Civilization: The Myth of the Peaceful Savage* (New York: Oxford University Press, 1996); R. E. Nisbett and D. Cohen, *Culture of Honor: The Psychology of Violence in the South* (New York: HarperCollins, 1996).

[23] It is widely accepted in the field of evolutionary science that differences between men and women "in physical development and physical competencies have almost certainly been shaped by sexual selection, and the majority of these differences have resulted from male-male competition over access to mates." See David G. Geary, *Male, Female: The Evolution of Sex Differences* (257). Geary is referring in particular to J. M. Tanner's entry on "Human Growth and Development" in *The Cambridge Encyclopedia of Human Evolution*, edited by S. Jones, R. Martin, and D. Philbeam (New York: Cambridge University Press, 1992), 98–105.

Especially interesting is Jonathan Gottschall's *The Rape of Troy: Evolution, Violence and the World of Homer* (2008), which emphasizes the key concept that Homer's heroes were driven primarily by sexual possessiveness as they strove to capture the most prized women and that their soldiers expected women as booty after a victory. Gottschall's method is informed by evolutionary psychology, and his work is associated with the field of literary Darwinism, as discussed below.

In Chapter 5, I place the heightened Victorian literary awareness of issues related to the ancient problem of male violence in the context of increased efforts to control it in the criminal justice system, as noted by historian Martin J. Wiener. The representation of various forms of male violence is central to the Victorian poetry studied here, and, along with other aspects of masculinity, it is emphasized in my discussions, but I want to make clear my critical assumption that male violence is a phenomenon that has always been characteristic of human experience. While I agree with Sussman and others that it is important to acknowledge "the multiplicity, the plurality of male gender formations" (Sussman, 8), I believe that linguistic and cultural constructions, despite their complexity, ultimately refer to the natural world and that what Sussman conceptualizes as fantasies about maleness as a "potent yet dangerous energy" are related to real-world phenomena far beyond the industrialization of nineteenth-century Britain and the growth of capitalism.

In a 2000 essay in *Victorian Studies* entitled "The End(s) of Masculinity Studies," Donald E. Hall approvingly refers to Angus McLaren's *The Trials of Masculinity: Policing Sexual Boundaries 1870–1930* (1997), which "traces the ways in which experts 'exploited the stereotype of a virile, heterosexual, and aggressive masculinity' to delimit male behavior in ways that met the needs of an industrial economy, a certain set of class interests, and their own specifically professional self-interest." He reads McLaren as opening up a new field of "transgender studies" that finally allows us "to disconnect gender behaviors from their supposedly 'natural' biological correspondents." (235). Hall may be correct in seeing "transgender studies" as a culmination of gender studies as conceived by the majority of gender theorists working in literary fields in the late twentieth and early twenty-first centuries, and I quote him here because of his special focus on Victorian literature. Unfortunately, this approach to gender studies, however full of insights about special class interests in a historical context, does not take advantage of, in fact systematically excludes, bodies of scientific knowledge about "the nature of human beings." However, there is hope for a more comprehensive, comparative, and realistic view in new critical approaches that could and should be applied to gender studies in a literary and cultural context.

When I generalize about human sex differences in the course of my discussions below, my generalizations are based on well-documented patterns in sexuality, parental behavior, and mating practices observed in recent times among human populations around the world, under the assumption that gender constructions in British Victorian society adjust to, conform with, and react against these universal patterns, which most likely have been dominant for at least a few thousand years.

Men are potentially able to procreate with large numbers of women, and the logic of evolutionary science tells us that those who procreated with the largest number of healthy, fertile women were the most likely to pass along their genetic heritage. In contrast, women have always had the incentive to be more highly selective and choose mates who would not only produce children but who had the necessary material resources and were willing and able physically to support their wives and offspring during the critical periods of pregnancy, childbearing, and childhood development.[24]

On the most basic level, Donald E. Brown's list of "human universals" characteristic of all known cultures, based on descriptions by ethnographers of observable phenomena (rather than deep or hidden structures or theoretical constructions) includes entries directly related to sex differences, such as "females do more direct childcare," "kin terms translatable by basic relations of procreation," "male and female and adult and child seen as having different natures," " males dominate public/political realm," males more aggressive," "males more prone to lethal violence," "sex (gender) terminology is fundamentally binary," "sex statuses," "sexual attraction," "sexual attractiveness," "sexual jealousy," "sexual modesty," "sexual regulation," "sexual regulation includes incest prevention," "sexuality as focus of interest."[25]

Literary Darwinism

In his essay "Getting It All Wrong," Brian Boyd offers a concise and penetrating critique of the prevailing view in academic literary studies that "the sciences, especially the life sciences, have no place in the study of the human world" (18). Boyd confronts the two fundamental claims of academic "Theory": *antifoundationalism*, that there is no secure basis for knowledge, and *difference*, that all assumptions about a supposedly universal human nature are "merely the product of local standards, often serving the vested interests of the status quo,"

---

[24] These generalizations are supported by virtually all major research related to sexual selection and human evolution, but they do not imply that individual men and women necessarily attempt to maximize procreation consciously. Rather, it is the case that individuals inherit traits that tend to lead in this direction. In addition to previously mentioned sources on human sexual differences, David M. Buss's *The Evolution of Desire: Strategies of Human Mating* (1994) is an easily accessible source of information, as is his *The Dangerous Passion: Why Jealousy is as Necessary as Love and Sex* (2000). In recent years, an increasing number of popular books on sex differences have appeared – for example, Steven E. Rhoads's *Taking Sex Differences Seriously* (2004) – and even on applications to literature: David P. Barash and Nanelle R. Barash's *Madame Bovary's Ovaries* (2005).

[25] See Brown's *Human Universals* (1991) and "Human Universals and Their Implications," in *Being Humans: Anthropological Universality and Particularity in Transdisciplinary Perspectives* (2000). A list of Brown's universals can be found in an appendix to Pinker's *Blank Slate*, 435–9.

claims that in effect deny human links to the natural world (20).[26] He argues that a biocultural approach, based on evolutionary epistemology, is, in opposition to academic Theory, characterized by "humility and hope" (24) in its search for truth while acknowledging "species-wide commonalities and differences" (29). I cite Boyd here because his essay explains and defends the critical assumptions of literary Darwinism that I have endorsed in this book in the context of a debate that has begun in academic departments in American and European universities and will probably continue for some time to come. My own work, as represented in this book, is not polemical, nor is it scientific. However, some in academia may be suspicious of any critical literary study that either refers to the "folk psychology" that enables individuals to process social information related to the self, other individuals, and social groups[27] or that attempts to integrate scientific scholarship with that of the humanities. Others might expect that a study like mine that cites the work of evolutionary psychologists and other scholars from scientific disciplines would itself employ scientific methodology in an attempt to obtain quantifiable results of some kind. In fact I have a deep interest in pioneering efforts in which literary scholars and social scientists collaborate to measure the responses of readers of literary texts in terms of universal sets of categories for analyzing meaning structures. Notably, a recent study by Joseph Carroll, Jonathan Gottschall, John A. Johnson, and Daniel J. Kruger, based on quantifiable data from questionnaires related to the reactions of readers to protagonists and antagonists in selected nineteenth-century British novels, has produced results that should be of interest to anyone studying the representation of human nature in literature.[28] I have long been fascinated by scientific efforts to study human responses to literature, but *Masculinity in Four Victorian Epics* is not a scientific work; it is a work of literary criticism.

The interpretation of literature, whatever its theoretical assumptions, is itself a distinctive scholarly genre. I attempt to explain my own assumptions about human nature, based not only on "folk psychology" or common sense or intuitions about the "nature of man" but, more formally, on a body of scientific scholarship that deals with the real world and that, like all science, is open to challenge by future research. In terms of human culture, language, and literature, our expanding understanding of human biology and psychology from an evolutionary perspective

---

[26] In *Literature, Science, and a New Humanities* (2008), Jonathan Gottschall outlines a trenchant and extensive criticism of the discipline of literary study in its current state: "the primary theoretical, methodological, and attitudinal struts that support the field are suffering pervasive rot" (3).

[27] See David C. Geary, *The Origin of Mind*, 11.

[28] "Human Nature in Nineteenth-Century British Novels: Doing the Math." Forthcoming in *Philosophy and Literature*. A book-length version of this study by Carroll, Gottschall, Kruger, and Johnson is under consideration: *Graphing Jane Austen: Human Nature in British Novels of the Nineteenth Century*. In his book *Literature, Science, and a New Humanities*, Gottschall makes an especially powerful argument for employing quantitative, scientific methods in literary study.

helps to account for our continuing interest in literary works from the historical past – like those of Tennyson, Barrett Browning, Clough, and Browning – as well as our capacity to imaginatively and ethically identify with human beings from around the world and from the most diverse linguistic, cultural, and national backgrounds.

I do not mean to imply that evolutionary theory has precisely mapped the development of the human mind or that all theorists working with "adaptationist" models agree on the precise structure of "human nature." However, in the recent past, evolutionary theory has begun to exert a powerful influence on the social sciences,[29] and now that influence is extending to the humanities as well. In the words of Boyd, "Science can explain why and how art has come to matter, but that will not give science the emotional impact of art, nor allow it to find a formula

---

[29] According to evolutionary theory, modern humans evolved from an ape-like creature during the Pleistocene age (from about 1.6 million to about 10,000 years before the present time, BP, when farming began) in an Environment of Evolutionary Adaptedness (EEA). As humans evolved in the EEA, certain behaviors and preferences resulted in reproductive success. Modern humans have existed for approximately 50,000 years – the first 40,000 as hunter–gatherers. The 10,000 years from the beginning of the Recent or Holocene age to the present is a relatively short period, and some argue that very little evolutionary change could have taken place. Although humans share some behaviors with all animals, a more narrow range of behaviors with all mammals, and a still narrower range with all primates – most of the behaviors that are considered distinctively "human" evolved during the Pleistocene in the EEA. See Richard G. Klein, "Archeology and the Evolution of Human Behavior" (2000), and Richard Potts, "Variability Selection in Hominid Evolution" (1998). Many scholars believe that social structure in the EEA was organized primarily around small, mobile bands of hunter–gatherers (from about 25 to about 150 individuals). Increasing intelligence was dependent on an increased time of dependency for offspring, and this in turn led to long-term bonding between males and females. See for example Robin Dunbar, "The Social Brain Hypothesis"; Hillard Kaplan, Kim Hill, Jane Lancaster, and A. Magdalena Hurtado, "A Theory of Human Life History Evolution: Diet, Intelligence, and Longevity" (2000); and Steven Jones, Robert Martin, and David Philbeam, *The Cambridge Encyclopedia of Human Evolution* (1992), 87, 464. As discussed above, literary Darwinist Joseph Carroll argues that what he terms the EP (Evolutionary Psychology) model, based on the idea that "the structure of the human mind at the present time and the structure of the mind of, say, half a million years ago, would be the same" ("Human Revolution," 34–5) is flawed and should be replaced by what he calls the EACA model, associated with the findings of evolutionary anthropology and cognitive archaeology, such as the work of William Irons, Robert Foley, and Rick Potts. The EACA model, as Carroll explains, takes into account the fact that in the past 100,000 years human beings "have evidently developed adaptive capacities and needs – intellectual, social, and cultural powers – of which their Paleolithic ancestors had no inkling" (40). Even if one accepts Carroll's critique, however, many of the findings of evolutionary psychologists remain relevant in the study of modern humans. A very useful and richly detailed introduction to the topic of evolutionary history and human genetics for the general reader is Nicholas Wade's *Before the Dawn: Recovering the Lost History of Our Ancestors* (2006). As Wade puts it, "Human nature is the set of adaptive behaviors that have evolved in the human genome for living in today's societies" (265).

for art, nor make art matter less. If anything, it will only clarify why and how art matters so much" ("Evolutionary Theories of Art," 172).

My point of view is consistent with the critical approach that has been called "literary Darwinism." Along with Boyd, Joseph Carroll is prominent among a group of literary scholars and critics who in recent years have been arguing for the integration of literary study with Darwinian social science. Carroll argues that "All the elements of the literary situation – the purposes of authors, the responses of audiences, the behavior, thought and feeling of [fictional] characters, and the formal properties of literary works – can be assessed and analyzed within the framework of adaptationist theory" (*Literary Darwinism*,164). This very large claim grows out of his fundamental assumption that "innate human dispositions exercise a powerful shaping on all forms of cultural order" (23), and he goes on to explain and illustrate his own highly developed method of describing the literary representations of interactions among innate dispositions, culture, and individual identity in a way that is consistent with adaptationist theory. Unlike those evolutionary psychologists who assume that the human mind as it exists today is still closely adapted to the hunter–gatherer way of life characteristic of the Pleistocene and who emphasize domain-specific aptitudes in humans, Carroll argues for the centrality of a general intelligence that facilitates an individual's flexible response to the environment and that drives the literary imagination. He cites a wealth of evidence from evolutionary specialists working in the fields of anthropology, cognitive science, and archeology to support the concept of a "distinctively human intelligence – an intelligence in which the powers of reflection, analysis, comparison, and creativity are uniquely developed" ("Human Revolution," 35). According to this view, "modern" human culture emerged in the course of the past 100,000 years, especially the past 50,000 years, and the evolution of human beings is a continuing process.

Before outlining the theoretical foundations of *literary Darwinism*, I want to point out that this general term is not used by everyone who takes a "biocultural" approach or applies adaptationist or evolutionary psychology to the study of literature. Boyd himself avoids the term because he is apprehensive that *Darwinism* implies a doctrinaire approach analogous to Freudianism and Marxism. Instead, he suggests the term *evolutionary literary criticism* or *evocriticism* (*On the Origin of Stories*, 388). I understand his desire to emphasize the fact that his commitment is to a research program, not a body of doctrine, and that research in the scientific fields of study initiated by Darwin or related to his work has gone beyond what Darwin himself could have foreseen. Nevertheless, I use the term *literary Darwinism* because in recent years it has been widely employed in both scholarly and popular accounts of studies that link literature to "human nature," and I am not sure that *evocriticism* will take its place. (Perhaps the earlier term *biopoetics*, as used in the title of the 1999 book edited by Brett Cooke and Frederick Turner – to which I refer below – will return to favor.) I join Boyd, as well as those like Carroll who have accepted the term *literary Darwinism*, in an open-minded embrace of future evidence relevant to literary studies from scientific disciplines open to experimentation and truth-testing.

Returning to a wider view of aesthetic issues, scholars influenced by adaptationist or evolutionary psychology have offered various theories to explain the function of art. For example, in *Homo Aestheticus: Where Art Comes From and Why* (1992) and *Art and Intimacy: How the Arts Began* (2000), Ellen Dissanayake argues that the arts are universally present in human societies because play and ritual were essential to the adaptation and survival of the species. Based on research in anthropology and ethology (the study of animal behavior, including that of humans), her thesis is that the arts have allowed humans to differentiate the special from the mundane, enabling them to cope with unusual situations and to gain a communal focus that enhances their ability to survive. More recently, in *The Art Instinct: Beauty, Pleasure, and Human Evolution* (2009), Denis Dutton offers a very useful overview of evidence supporting claims that the production and appreciation of art arises from universal, innate dispositions. Dutton's work is especially significant in explaining how these dispositions can be explained primarily in terms of sexual (rather than natural) selection.

Such views account for the basic function of art prior even to the oral antecedents of written words. Literary Darwinists acknowledge such theories but of course focus on the function of literature itself. Carroll himself suggests that the arts "serve a vital adaptive function – that of organizing human motives and thus ultimately regulating behavior," and he emphasizes the "detached self-consciousness" that requires modern humans to "live in and through their own imaginative structures" ("Human Revolution," 41). For him, "[t]he distinguishing characteristic of literature is that it creates an imaginative order in which simulated experience can take place" (44). The individual author's "beliefs, values, and attitudes" (his or her "point of view") are vital to the study of any literary work of literature, and "[a]ll individual identities are shaped partly by innate characteristics – the elements of human nature that vary within the range of individual differences – and partly by the conditions of experience" (45). "Experience" here can be both individual and collective ("cultural"). Carroll has been a pioneer in integrating the theory of adaptationist psychology with literary studies, beginning with his book *Evolution and Literary Theory* (1995), and in a recent article, he offers a most comprehensive and up-to-date discussion of "an evolutionary paradigm for literary study" (Carroll, "Evolutionary Paradigm"). Other critics associated with a Darwinian, "adaptationist," "biopoetic," or "biocultural" perspective include the previously mentioned Boyd as well as Brett Cooke, Robin Fox, Jonathan Gottschall, Margaret Nesse, Michelle Sugiyama, Ian Jobling, Eric S. Rabkin, David Sloan Wilson, Carl P. Simon, Nancy Easterlin, Robert Storey, and Frederick Turner, among others.[30]

---

[30] The 2008 special issue of *Style* that features Carroll's "Evolutionary Paradigm" article includes "responses" by Brian Boyd, Gorden M. Burghardt, Brett Cooke, Frederick Crews, Ellen Dissanayake, Karl Eibl and Katja Mellmann, Lylle Eslinger, Jeffrey E. Foy and Richard J. Gerrig, Harold Fromm, Eugene Goodheart, Jonathan Gottschall, Torben Grodal, Geoffrey Galt Harpham, Patrick Colm Hogan, Tim Horvath, Tony Jackson, Fortis Jannidis,

In his 2004 book *Literary Darwinisn: Evolution, Human Nature, and Literature* Carroll acknowledges that he and the other literary Darwinists do not have a "full and adequate" concept of human nature and points out complexities such as the distinction between "'ultimate' regulative principles of inclusive fitness or reproductive success" and "'proximal' mechanisms that operate on the level of immediate triggers to behavior" (xviii), but he provides a working model that can be applied to the interpretation of literary texts and illustrates his theoretical approach with discussions of Jane Austen's *Pride and Prejudice* and other well-known works that add to our critical understanding. In addition, Boyd's 2005 essay "Literature and Evolution: A Bio-Cultural Approach" provides a helpful introduction to the field as a whole, and his important recent book *On the Origin of Stories* (2009) incorporates work from several earlier publications and offers a comprehensive study of the application of evolutionary psychology and models of human nature to literature. Also of considerable interest are the essays (including ones by Carroll and Boyd) collected in *The Literary Animal*, edited by Gottschall and Wilson (2005), and the earlier *Biopoetics: Evolutionary Explorations in the Arts* (1999), edited by Cooke and Turner.[31]

Carroll's emphasis on "point of view" is related to parallel developments in "cognitive literary theory," which also draws on evolutionary psychology and the function of the human brain in writing, reading, and interpreting literature. Lisa Zunshine's *Why We Read Fiction: Theory of Mind and the Novel* (2006) is particularly interesting in this context. The term *theory of mind* refers to the individual's ability to "read the minds" of others, and this is central to the experience of reading and interpreting fictional narratives. In her discussion of a wide range of texts, she combines the insights about the "mind-reading" that goes on in the reading process – as readers not only use but experiment with their evolved cognitive adaptations – with "metarepresentationality," the cognitive ability to keep track of representations of characters' thoughts and mental states in order to conceptualize and construct the narrative as a whole. It seems to me that "theory of mind" helps to account for the popularity (among teachers of literature) of Wayne C. Booth's 1961 book *The Rhetoric of Fiction*, with its key ideas of "the implied author," "the unreliable narrator," and so on.

Frank Kelleter, Amy Mallory-Kani and Kenneth Womack, David S. Miall, David Michelson, Catherine Salmon, Judith P. Saunders, Michelle Scalise Sugiyama, Roger Salmon, Edward Slingerland, David Livingstone Smith, Murray Smith, Ellen Spolsky, Robert Storey, Peter Swirski, and Blakey Vermeule. Earlier, a special issue of of *Philosophy and Literature* (October 2001), edited by Nancy Easterlin, contains relevant essays by Brian Boyd, Lisa Zunshine, Michelle Scalise Sugiyama, Easterlin, and Jonathan Gottschall.

[31] Among other recent publications related to theoretical considerations of evolution and literary studies are Marcus Nordlund, "Consilient Literary Interpretation" (2002); Jonathan Gottschall, "The Tree of Knowledge and Darwinian Literary Studies" (2003); and Edward Slingerland, *What Science Offers the Humanities: Integrating Body and Culture* (2008).

Patrick Colm Hogan's *The Mind and Its Stories: Narrative Universals and Human Emotion* (2003) is another important study that applies cognitive science to the study of literature in ways that are helpful not only to theorists but to critics who analyze and interpret individual works. Like the literary Darwinists, he acknowledges the concepts of "human nature" and "literary universals," and his own study of non-European literatures, in particular classic Indian and Chinese works, has convinced him that certain basic emotions are associated with literary narratives from around the world. Furthermore, the prototypes of "emotion terms" provide the basis for prototypical narratives or stories. Prototypes for genres, character types, and scenes guide our interpretations of the stories. In prototypical stories, "the complex process of narrative construction is guided and organized by the expansion or elaboration of the micronarratives that define the prototype eliciting conditions for happiness, whether or not those conditions are ultimately achieved in the narrative" (118–21). Of course, some stories end in sorrow rather than happiness, but Hogan interprets the sorrow-based structure as a "truncated" form of the happiness-based structure. The "predominant prototypes for happiness are (a) romantic union with one's beloved and (b) the achievement of political and social power, both by an individual and by that individual's in-group (for example, his or her nation)." The "happiness prototypes define romantic and heroic tragic–comedy, which are the most common and most prominent narrative structures cross-culturally" (121). Among other critics in the field of cognitive literary theory today are Mark Turner and Alan Palmer.[32]

Closely related to cognitive literary theory and literary Darwinism is "ecocriticism," focusing on literature in which the relationship between humans and the environment plays a central role. Easterlin, for example, listed above with other critics who have demonstrated a Darwinian point of view in their work, has shown a strong interest in ecocriticism.[33] Also, Glen A. Love, in *Practical Ecocriticism: Literature, Biology, and the Environment* (2003), draws on the work of Carroll and other literary Darwinists in championing the scientific method and gently but firmly distancing himself from a "nature-endorsing postmodernism": "We require the standards of evidence and rational thought to move us beyond

---

[32] Useful introductory works to the field of cognitive literary studies or cognitive poetics include Peter Stockwell, *Cognitive Poetics: An Introduction* (2002); Gilles Fauconnier and Mark Turner, *The Way We Think: Conceptual Blending and the Mind's Hidden Complexities* (2002); and Joanna Gavins and Gerald Steen, eds, *Cognitive Poetics in Practice* (2003).

[33] Easterlin has explored a variety of other "biocultural" approaches as well. Her book manuscript, *What is Literature For? Biocultural Theory and Interpretation* (under consideration), extending the insights in her 2005 article "How to Write the Great Darwinian Novel," demonstrates the value of cognitive and evolutionary theories of the arts while emphasizing her view that, although these scientific theories are clearly relevant to literature and very helpful to critics, scientific disciplines, based on experimentation and quantitative analysis, are fundamentally different from literary and aesthetic criticism, with its emphasis on formal analysis and interpretation.

attractive theories of unreality" (45). Love strongly endorses the work of E. O. Wilson and emphasizes his concept of "*biophilia* – reverence for life, our instinctive sense of ourselves as creatures of natural origins" (70). His fundamental assumptions about literary criticism are compatible with those of Dissanayake, who in *Homo Aestheticus*, cited above, finds the origin of all art forms in humans' sense of wonder in experiencing the natural world. Like Hogan, Love stresses the emotional basis of all literature, but he is primarily concerned with texts in which some aspects of the natural environment are prominent. Love is a specialist in American literature (although he makes references to the ancient Greek and Roman genre of the pastoral), and the primary texts analyzed in *Practical Ecocriticism* are by Willa Cather, Ernest Hemingway, and William Dean Howells. Nevertheless, he, like other critics discussed in this chapter, is working with human predispositions associated with the universals of human nature, and his fundamental approach to an author's relationship with the natural world might be applied to the Spasmodic Smith, for example, who, as pointed out earlier, has his protagonist Walter record his spontaneous, emotional reactions to sunsets, night skies, and other natural phenomena as poetry and develop his romantic relationships in the context of prominent landscape settings.

Moving to Tennyson, Barrett Browning, Clough, and Browning, I refer to some of the critical works cited above in my discussions of the long poems. Also, I make use of Carroll's model for analyzing literary narratives.[34] He begins by outlining a "hierarchical motivational structure of human nature" beginning with "inclusive fitness" (survival, development, reproduction – not necessarily conscious motives in the fictional characters) as "the ultimate regulative principle." Below that he lists seven "behavioral systems" that organize the "somatic" and "reproductive" efforts that motivate and activate the emotions of humans: survival, technology, mating, parenting, kin relations, social relations, and cognitive activity (including literature and the other arts). Each behavioral system or pattern incorporates various "cognitive modules." The universals of Darwinist theory are "merely behavioral patterns so firmly grounded in the logic of human life history that they are characteristic features of all known cultures" (91). Obviously, the "mating" category in particular accounts for the primary elemental motives and organizing principles of many (perhaps most) human narratives. The behavioral patterns that define human nature in Carroll's model are assumed to interact with cultural and individual differences. Individual authors may of course resist what they perceive to be the conventions of the culture in which they live, and they may not ground their own values in human nature, but the tendencies and patterns of universal human nature provide the background, the context for even stories by highly unconventional authors or about highly individualistic characters – like Claude, Clough's protagonist in *Amours de Voyage*, for example. Carroll's emphasis on

---

[34] Carroll describes his methodology in detail in his essay "Human Nature and Literary Meaning: A Theoretical Model Illustrated with a Critique of *Pride and Prejudice*" (2005), from which I quote below.

human emotions and narratives based on primary behavioral systems can be usefully compared with Hogan's prototypes.

The approach of the literary Darwinists may still be controversial (or unknown or misunderstood) among the majority of literary theorists today; however, it is not only consistent with the concept of evolutionary science and the recent findings of adaptationist psychologists and others as discussed, but also helps to explain the cogency of traditional, "humanist" concepts of human nature as well as the force of formerly influential literary and cultural theories by Joseph Campbell, Northrop Frye, and others who postulated "archetypal" human universals. It is also in line with the findings of anthropologists like Donald E. Brown who "take it as axiomatic that human universals must play a part in understanding what it is to be human" (172). It is interesting that evolutionary psychologists themselves have been accused of "essentialism" in their developing concept of "human nature." For example, David J. Buller thinks that "human nature is just as great a superstition as the creation myth of natural theologians" (480). Within the scientific community itself, there is controversy over issues such as the evolutionary psychologists' emphasis on domain-specific aptitudes in humans, but as E. O. Wilson writes in his Foreword to *The Literary Animal*, "Science is neither a philosophy nor an ideology. It is a way of exploring the tangible world that conferred understanding and power beyond the imaginings of prescientific people" (x). It is reasonable for literary scholars to take advantage of scientific findings about humans and the natural world because that is what literature is about. This does not mean that literary criticism must become a scientific discipline, although studies that combine literary and scientific methodologies (as, for example, in models of how the brain functions during the reading process or statistical evaluations of readers' responses to a particular text) are of great potential. As already noted, I do not intend to suggest that my own work is "scientific," only that I profit from surveying the work of scientists in fields related to human nature, as discussed in this Introduction. One source, not yet mentioned, that has been especially helpful to me is David C. Geary's comprehensive study of the human mind, *The Origin of Mind* (2005),[35] an interdisciplinary, Darwinian synthesis of research from genetics, neuroscience, adaptationist psychology, cognitive science, and related fields. Geary includes, for example, an in-depth study of *general intelligence*, a concept central to Carroll's version of literary Darwinism, as noted earlier, and an account of *theory of mind*, the ability to "read minds" and make inferences about

---

[35] Geary is a psychologist who does not himself suggest "literary" applications of his work. He says of *The Origin of Mind*: "I worked under the assumption that motivational, affective, behavioral, cognitive, and brain systems have evolved to process social and ecological information patterns ... that covaried with survival or reproductive options during human evolution. My specific proposal is that all of these systems are ultimately and proximately focused on supporting attempts by the individual to gain access to and control of the social (e.g., mates), biological (e.g., food), and physical (e.g. demarcation of territory) resources that supported survival and improved reproductive prospects during human evolutionary history" (xiix).

the intentions of other people that Zunshine finds crucial to our ability to read and interpret literary texts. It is especially significant to me that although Geary is not often cited by literary Darwinists and in his own work he has little to say about literature *per se*, his account of human psychology is largely consistent with that used by Carroll, Boyd, and like-minded literary theorists. Earlier works by Geary, including his book *Male, Female* (1998) and subsequent articles,[36] have special relevance in the study of human sexual differences, and many of his findings in that area have been incorporated into *Origin*.

To return to the scene of Darwinian dilemmas in the Victorian age, Arnold, in one of his best known critical essays, made large claims for the study of literature by appealing to "the constitution of human nature," which he believes is "built up" by the "powers" of "conduct," "intellect and knowledge," "beauty," and "social life and manners."[37] Ironically, his emphasis on human nature is in the context of his debate with Darwinist T. H. Huxley about the relative emphasis on literature and science in university curricula. For Arnold, the study of literature relates knowledge about the physical world to human psychological needs, and versions of this claim are still viable today. I think it is likely that the influence of "literary Darwinism" will increase in the future as common misconceptions about its negative ideological and political implications fade. Prominent among these is the assumption that any version of "human nature" based on evolutionary psychology implies "genetic determinism" – one form of which would be the "essentialist" concept of the male rejected by Sussman as discussed above. Cognitive theorist William C. Dennet gives an extended refutation of the "genetic determinism" argument in *Freedom Evolves* (2003). Countering the claims of Stephen Jay Gould and other critics, Dennet argues that "The issue is not about determinism, either genetic or environmental or both together; the issue is about *what we can change* whether or not our world is deterministic" (160). Boyd agrees that a "bio-cultural" approach does not imply determinism: "Genes do not *constrain*; they enable ... Without genes, flexibility of behavior, culture and learning would all be impossible" ("Literature and Evolution," 3). Tennyson, Barrett Browning, Clough, and Browning – like many of their readers through the years – celebrated certain aspects of human nature but yearned for change as well. In each case, implicit or explicit desires for social change and reform are associated with Christian idealism. This is all consistent with a broadly conceived definition of literary art in the context of literary Darwinism. As Carroll puts it, "We can define art as the disposition for

---

[36] See especially "Sexual Selection and Human Life History" (2002) and "Sexual Selection and Sex Differences in Human Cognition" (2002).

[37] "Literature and Science," in which Arnold answered T. H. Huxley's "Science and Culture" lecture given at the opening of the Scientific College at Birmingham in 1880, was delivered as the Rede lecture at Cambridge in June 1882 and then became the most popular lecture in Arnold's tour of the United States in 1883–84. Carroll (2004) points out that even the cultural values of Arnold that have "an absolutist and transcendental character ... can be assimilated to a relativistic Darwinian model of cultural values" (4).

creating artifacts that are emotionally charged and aesthetically shaped in such a way that they evoke or depict subjective, qualitative sensations, images or ideas. Literature, specifically, produces subjectively modulated images of the world and of our experience in the world" ("Evolutionary Paradigm," 122).

Now that I have outlined my theoretical and critical assumptions, I want to add a few points about my methods before proceeding to the chapter on Tennyson's *Idylls*. The chapters that focus on individual authors and works are meant to be read sequentially rather than as freestanding essays. For example, observations about the concept of chivalry in the Tennyson chapter will anticipate my treatment of that topic in subsequent chapters. As already implied, an extended discussion of representations of masculinity in the four poems invites reconsideration of a wide range of thematic as well as aesthetic and formal issues – from the sometimes complex positions of the individual poets on social, philosophical, and religious issues and their ideas about the genre of the long poem to theoretical questions about literary explorations of human nature. The concepts of literary Darwinism encourage a close reading of a literary text: attention to literary devices and images calculated to elicit spontaneous emotional responses from the reader and in narratives the structure of plots that make sense in terms of human life stories. More controversially, they encourage the postulation of a real author. Literary texts are written and read by real people, and even in the case of anonymous or disputed authorship, readers have the predisposition to speculate about the author as a real person. As generalized by Dutton, "it is from an evolutionary standpoint psychologically impossible to ignore the potential skill, craft, talent, or genius revealed in speech and writing" (176).

And yet for more than a half-century, the desirability of de-emphasizing or eliminating consideration of the author's intentions from evaluation of literary texts in academic literary criticism has been dominant. This principle is implicit in the "intentional fallacy" associated with New Criticism, postmodernist assumptions about the "death of the author," and extreme versions of historicism and "social construction," as discussed above. My primary purpose is not to attack or refute the critics associated with these traditions but rather offer an alternative approach in accord with evolutionary psychology and the emerging paradigm of the literary Darwinists. In the case of each of the major Victorian poets studied, I will begin with an overview of the poet's career and the place of the individual work under consideration in that career. Available biographical and historical information helps us to understand the author's self-concept as an artist, his or her developing literary career, and relationship with readers in a continuing attempt to capture their attention as conscious, curious, individual human beings and achieve their respect while simultaneously appealing to communities of readers in terms of nationalism or social group identity and religious or other broad or universal value systems. In reading the individual works, it is appropriate not only to point out especially striking or representative passages for analysis but also to refer to the overall structure of the narratives. Occasionally "plot summaries" are appropriate, not merely to remind readers of the "story" of the poem but to demonstrate how that story is organized in terms of the lives of characters; contextualized within particular cultural traditions; and, in a larger sense, within cross-cultural, universal patterns associated with human nature. Of course in dealing with human nature and with related social structures, large and small, I emphasize male characters and issues of masculinity.

Chapter 2
Tennyson's Arthur and Manly Codes of Behavior

Beginning with Tennyson's contemporaries, one of the targets for hostile criticism of *Idylls of the King* was his adaptation of the warrior–king protagonist portrayed in Malory's *Le Morte D'arthur*. Swinburne mockingly referred to the "morte d'Albert, or Idylls of the Prince Consort" and proclaimed that Tennyson had "lowered the note and deformed the outline of the Arthurian story, by reducing Arthur to the level of a wittol, Guinevere to the level of a woman of intrigue, and Launcelot to the level of a 'co-respondent.'"[1] Henry Crabb Robinson thought Tennyson's Arthur was "unfit to be an epic-hero" (II:792), and Henry James called him a prig (177). T. S. Eliot asserted that Tennyson had adapted "this great epic material – in Malory's handling hearty, outspoken and magnificent – to suitable reading for a girls' school."[2] Clearly, gender is central to all these comments – contemporary readers were dissatisfied with the way Tennyson portrayed Arthur's "manhood," a problem of which Tennyson was well aware, and many modern readers either voice concerns similar to those of Swinburne and T. S. Eliot or offer readings of the *Idylls,* which in my view seriously distort Tennyson's attempt to explore gender issues. Tennyson's modern editor and interpreter Christopher Ricks makes no serious attempt to defend Tennyson against the traditional critics who ridicule the Poet Laureate for making Arthur a wimp, something less than a real man.[3] On the other hand, more politically attuned critics taking a feminist or gender studies approach either ignore or deal inadequately with the question of Arthur's masculinity. Especially disappointing is Elliot L. Gilbert's 1983 *PMLA* essay, "The Female King: Tennyson's Arthurian Apocalypse," probably the most influential study of its kind.[4]

As is well known, Tennyson was fascinated by the King Arthur legend from the beginning of his poetic career. Hallam Tennyson's reference to his father's memories about the "vision of Arthur" that had come upon him "when, little more

---

[1] See "A. C. Swinburne Replies to Taine," 339; "A. C. Swinburne on the *Idylls*," 319.

[2] Quoted by Christopher Ricks in his edition of Tennyson's poems, 258.

[3] Ricks's critique of the poem is of course primarily an aesthetic and formal one, and he places a generally low estimate on the *Idylls*, relative to Tennyson's work as a whole. "No other poem of Tennyson's was created with such a central uncertainty as to its shape, style, sequence, and size. Such uncertainty in composition is not in itself any evidence of ultimate uncertainty in achievement. Yet *Idylls of the King* must be judged strikingly uneven …" (250).

[4] Sussman refers to Gilbert's essay as "incisive and influential" (208), and Shaw incorporates Gilbert's main ideas into her feminist reading of the *Idylls* (92).

than a boy, I first lighted upon Malory" (II:128) has been circulated by generations of critics. However, Roger Simpson has shown in detail that multiple sources of the Arthurian legend were available to Tennyson prior to his first formal use of it in his poetry in the early 1830s.[5] Tennyson's 1830 volume, *Poems, Chiefly Lyrical*, published when he was 21 years old, includes a fragment entitled "Sir Lancelot and Queen Guinevere," followed by "The Lady of Shalott" in 1832. Unpublished Arthurian items from the early 1830s include notes from Collinson's *History of Somersetshire*, a prose sketch entitled "On the latest limit of the West," a memorandum with the heading "K. A. Religious Faith" (later presented to James Knowles), a "five-act scenario," an outline of the early books of Malory's *Le Morte D'arthur*, and additional Lancelot fragments written with a projected ballad in mind (Simpson 190). Although his notes from Collinson indicate at least some interest in a historical Arthur early on, one of the most interesting things about Tennyson's earliest Arthur poems is their lack of historical and topographical specificity. Only later, in the late 1840s, while actively engaged in the composition of the *Idylls* as a whole, did he visit historical sites in Cornwall and Wales to get a firsthand impression of the supposed Arthurian topography.

Clearly, Tennyson was interested in Arthur primarily as a mythic, not as a historical, figure. In his "Palace of Art" (1832), for example, we find "mythic Uther's deeply-wounded son / ... dozing in the vale of Avalon, / And watched by weeping queens" (105–8). Many critics have seen Arthur as a Christ figure.[6] At the same time, however, Tennyson wished to represent Arthur not as supernatural hero but as a *man* who had affinities with his dead friend Arthur Hallam (and with the "uncrowned king," Prince Albert, as the Dedication of 1862 makes clear). In introducing *The Holy Grail and Other Poems* (1869), Tennyson wrote that Arthur is "meant to be a man who spent himself in the cause of honour, duty and self-sacrifice, who felt and aspired with his nobler knights, though with a stronger and clearer conscience ... God had not made since Adam was, the man more perfect than Arthur." And the 1891 addition of the phrase "Ideal manhood closed in real

---

[5]   See Roger Simpson, *Camelot Regained: The Arthurian Revival and Tennyson, 1800-1849*, 1–4.

[6]   In *The Fall of Camelot: A Study of Tennyson's Idylls of the King* (1973), John D. Rosenberg is particularly astute in interpreting Tennyson's dilemma in representing the paradoxical Christ-like duality of Arthur, who is both human and divine. For example, in "'Guinevere' ... for an excruciating instant we catch a glimpse of Christ with horns" (127). As a "cuckolded husband he cannot speak like a surrogate god," but "[i]t is worse still when he tries to speak like a 'real man'" (130). Rosenberg also argues that "[a]lthough the King is a Christ figure in origins, mission, and promise of his return, he is also a solar deity ... Arthur is so closely linked to the sun throughout the *Idylls* that his character never wholly detaches itself from the symbol, or the symbol from that ancient body of belief in which the gods, once housed in the heavens, descend to earth, are worshipped as heroes, and fructify the land" (42). In discussing possible connections between Tennyson and the ideas about myth held by the Rev. George Stanley Faber, W. D. Paden similarly combines the concepts of Arthur as a warrior of God and his identification with a pagan sun deity (79–88).

man" describing Arthur in the epilogue "To the Queen" was, according to Hallam Tennyson, the last correction to the *Idylls*, made because "my father thought that perhaps he had not made the real humanity of the King sufficiently clear" (II:129). In terms of Hogan's "narrative universals," as discussed in Chapter 1, Tennyson, on one hand, was fascinated with the heroic, tragic Arthurian myth focusing on the achievement of political and social power by Arthur and his order of the Round Table followed by his downfall and, on the other, offered a degree of literary realism in his portrayal of Arthur as a leader and the dysfunctional "romantic union" between Arthur and Guinevere in the context of this archetypal narrative involving sex and procreation.

"Morte d'Arthur" (1842) was the first published poem that would be incorporated into the *Idylls* (although the later "Lancelot and Elaine" is to some extent a retelling of "The Lady of Shalott," originally published in 1832). However, "Morte d'Arthur" had been written during the period 1833–34, like several other key poems in Tennyson's oeuvre, "under the shock" of his friend Hallam's death, and it is easy to see a connection between the Christ-like Arthur and the Hallam of *In Memoriam*, who, like Jesus, exhibited a manhood combined with female grace.[7] It would be 15 years after the publication of "Morte d'Arthur" before six copies of a "trial volume" entitled *Enid and Nimuë: The True and the False* were printed (Nimuë was the first name given to Vivien) – and the first volume to bear the title *Idylls of the King*, incorporating "Enid," "Vivien," "Elaine," and "Guinevere" – appeared in 1859. The "Dedication" to Prince Albert was added in 1862.

Beginning in late 1869, the publication process gathered momentum: in that year appeared "The Coming of Arthur," "The Holy Grail," "Pelleas and Ettarre," and "The Passing of Arthur" (the original "Morte" with additions), and in the next year came three expanded versions of the 1859 *Idylls*: "Geraint and Enid," "Merlin and Vivien," and "Lancelot and Elaine." "Gareth and Lynette" and "The Last Tournament" (the latter published individually the previous year in a journal) were published together in 1872, and the Imperial Edition of Tennyson's works (1872–73) brought together all of the *Idylls* except "Balin and Balan" and added a new epilogue: "To the Queen." "Balin and Balan" appeared as part of another collection in 1885, and in 1889, three years before Tennyson's death, the complete *Idylls* was published with the poems in their final order: "Dedication," "The Coming of Arthur," "Gareth and Lynette," "The Marriage of Geraint," "Geraint and Enid," "Balin and Balan," "Merlin and Vivien," "Lancelot and Elaine," "The Holy Grail," "Pelleas and Ettarre," "The Last Tournament," "Guinevere," "The Passing of Arthur," and "To the Queen."

Although Tennyson wrote most of the material finally incorporated into this final version during two relatively intense phases of creativity, 1856–59 and 1868–74, it is fair to say that he was enthralled by the Arthurian subject throughout his career, and that in a sense the Arthurian project dominated his whole career. Early in the process, Tennyson was interested in exploring issues of creativity

---

[7] Hallam Tennyson quotes his father on "the man-woman" in Christ, the union of tenderness and strength (H. Tennyson I: 326n).

and the relation of the artist to the world in the richly symbolic (but not strictly allegorical) story of the Lady of Shalott. Also very early he was drawn to the story of the adulterous Lancelot and Guinevere and then a little later to contrasting stories of "true and false" ladies. However, the figure of Arthur was always central to the *Idylls* project: though he is in a formal sense the protagonist only in the first and last idylls, his presence dominates the entire series and gives a unified meaning to the story as a whole. As John D. Rosenberg observes, Tennyson's original "Morte d'Arthur," drafted in 1833, revised in 1835, and published in 1842, was the "germ of the whole ... instinctually right in tone and design" (13) and, with the initial framing section added in 1869, remained in this form to the end, though Tennyson continually expanded and substantially revised the larger poem over the decades. In a sense, the ending is implicit in the beginning of any narrative, but Tennyson literally wrote the ending first. At the beginning of "Morte d'Arthur," the great catastrophe of the final battle already has taken place – Arthur's Order of the Round Table has been definitively destroyed, and nearly all of the knights are dead (an absent, repentant Lancelot apparently lives on to become a holy man, and Bedivere survives in the role of storyteller). The additional 169 lines added at the beginning when "Morte" became "The Passing of Arthur" in 1869 supply a description of the last bloody one-on-one encounter between Arthur and Modred and intensify the original effect of desolation.

In this chapter I consider the *Idylls* as a unified whole, referring to the final, complete version first published in 1889, with the exception of a few instances in which the order of composition may help to clarify a point. However, I want to point out that, beyond the intensified images of death and destruction added to the final idyll, the general tendency in Tennyson's revisions and additions through the years was to place increasing stress on patterns of violence and the issue of "manhood." The "Geraint and Enid" story, amplified and divided into two individual idylls in 1873 (part one was renamed "The Marriage of Geraint" in 1886), and "Balin and Balan," the last-written idyll, originally published in 1885, are focused, in the first instance, on an extreme case of masculine gender identification and its relation to formulaic violence, and, in the second, on the fear of an innate male capacity for irrational violence and its relation to madness.

Gilbert's description of Arthur as a "female king" and his contention that Tennyson, preoccupied with the idea of "woman as cosmic destructive principle,"[8] meant to show in the *Idylls* that Arthur's ideal community is destroyed by "an irrepressible female libidinousness ... released by the withdrawal of patrilineal authority" (873), is based on a confused reading of the poem that somehow misses Tennyson's focus on the problematics of male sexuality and capacity for irrational violence. Most obviously, there is nothing in the poem to suggest that Arthur is not a male (unless he is a supernatural god figure in the form of a man). It is his gender, not his sex, that has seemed problematical to some readers. From the beginning of his career, Tennyson was sensitive to his own ambiguous social status as a male poet in the Romantic tradition, associated with the suspiciously feminized qualities

---

[8] Gilbert quotes Gerhard Joseph on Tennyson's notion of woman as cosmic destructive principle (873).

of imaginative inwardness, emotive openness, isolation from the aggressive "entrepreneurial manhood" valorized by bourgeois ideals.[9] He experimented with representations of androgyny throughout his career, most notably in his long poem "The Princess" (1847), but even there, the essential difference between men and women is finally upheld: "For woman is not undevelopt man, / But diverse ... / Not like to like, but like in difference" (VII.ll.259–62). This story of a princess who founded an all-female college for the purpose of emancipating women is described by Marion Shaw as the "most comprehensive" of the literary works dealing with the "women-and-marriage question" at midcentury. However, Shaw also comments on Tennyson's "acute anxiety concerning male sexual needs and definitions" and refers to Kate Millet's observation that the poem "takes fright at its own daring and turns away from the logical pursuit of its argument" for fear of the consequences of women's independence (42). It seems to me that the best explanation for Tennyson's drawing back from androgyny as an ideal is not a generalized male fear of the "new woman" but a recognition of the source of what he thought to be insurmountable differences between men and women: male violence and the exclusively male role of the warrior. For Tennyson it is inconceivable that Princess Ida and her (otherwise Amazonian) followers should actually arm themselves and physically defend their College against the Prince and his warriors. (However "feminized" the cross-dressing Prince may be, he is still a warrior and is expected to submit himself to physical danger and possible death on the battlefield.) The Prince's militaristic father may be a patriarchal ogre (even to Tennyson), but he understands this essential difference.[10] The women are of course represented by male mercenaries in the staged battle of evenly matched forces. It is no accident that the Princess is overcome with sympathy for the wounded Prince and falls in love with him as she nurses him back to health.

For Tennyson, the image of a wounded warrior under the care of his lover–nurse is laden with symbolism: it is the ultimate representation of tenderness and romantic love between the sexes, and it is also the ultimate representation of gender difference. In the *Idylls* we see it most memorably in Elaine's unequivocal but hopeless love for the wounded Lancelot, who cannot fulfill his proper role as knightly lover to his loving nurse because of his improper commitment to the Queen. The noble but unfairly suspicious Geraint proudly keeps aloof from his

---

[9] Sussman writes, "Entrepreneurial manhood with its emphasis on engagement in the male sphere of work, its valuing of strength and energy, and its criterion of commercial success measured by support of a domestic establishment generated particularly acute anxiety for the early Victorian male poet, for this definition of male identity conflicted with the ideal of the poet based on a romantic model in many ways constructed to oppose the new formation of bourgeois man" (82).

[10] An interesting late Victorian treatment of this theme is Walter Besant's novel *The Revolt of Man* (1882), set in a future when women are in control and men are subordinate. When the men stage a revolt and form a rebel army, the female authorities empty the prisons in order to build an opposing army of male soldiers. However, a group of women sympathetic to the men's movement frighten away the disorganized and ignoble conscripts before any fighting can take place.

wife Enid as he kills and mutilates one challenger after another until, after secretly bleeding under his armor, he falls from his horse and she tenderly swaths his wound with the veil of faded silk she has torn off her face. Later, after Geraint had been transported to the castle of Earl Doorm and has awakened to find "his own dear bride propping his head, / Chaffing his faint hands, and calling to him; / And felt the warm tears falling on his face" ("Geraint and Enid," 583–5), he continues to feign sleep in order to savor her display of devotion. Significantly, in "The Passing of Arthur," the (supposedly) dying Arthur is not comforted by his unfaithful wife, who is far away in a convent, but by the fantasy queens who will accompany him on his journey to Avalon. In any case, the ultimate gender reversal that would be represented by, let us say, the image of a wounded female warrior, nursed by the man for whom she has fought and is perhaps dying, is simply unimaginable for Tennyson or any other Victorian writer.[11]

Among the most fundamental gender distinctions implicit in the *Idylls* are those associated with the "violence against women" taboo enforced by the code of chivalry. The absolute nature of this taboo is represented in the "Geraint and Enid" idyll when the wounded Geraint, whose condition has continued to deteriorate so that he is presumed to be dying or already dead, is miraculously revived by the bitter cry of Enid when the Earl, frustrated with Enid's failure to respond to his advances, "unknightly with flat hand, / *However lightly*, smote her on the cheek" (717, emphasis mine):

> This heard Geraint, and grasping at this sword
> (It lay beside him in the hollow shield),
> Made but a single bound, and with a sweep of it
> Shore thro' the swarthy neck, and like a ball
> The russet-bearded head roll'd on the floor.
> So died Earl Doorm by him he counted dead. (724–9)

In "Pelleas and Ettarre," Gawain, casual about most of his knightly vows but to whom violence against women is unthinkable, advises his younger and more idealistic colleague on the way to handle the servants sent by Ettarre to restrain her persistent lover:

> "Why, let my lady bind me if she will,
> And let my lady beat me if she will:
> But an she send her delegate to thrall
> These fighting hands of mine – Christ kill me then
> But I will slice him handless by the wrist,
> And let my lady sear the stump for him,
> Howl as he may." (325–32)

---

[11] Less obviously, but just as surely, it is absurd to imagine Victorian girls joining Robert Baden-Powell's (late Victorian) Boy Scouts, who are in effect being groomed to fight and possibly die in war. Scouts' rules were to a large measure derived from chivalric codes, and recommended reading-lists for Scouts were dominated by books on chivalry and King Arthur. See Mark Girouard, *The Return to Camelot*, 255.

Because manhood in the *Idylls* is defined primarily in terms of the warrior role, and the most fundamental duty of the knightly warrior is to protect women, a female knight is inconceivable. It is not primarily the fear of an armed "Joan of Arc" figure usurping the role of the male warrior but rather the "violence against women" taboo and the fear of the mutilation and death of women on the battlefield that accounts for the sharply defined gender boundaries here. A chivalric code emphasizing gallantry toward and protection of women, along with the traditional virtues of loyalty, valor, mercy, modesty, friendship, and honor, is most basic among the ideals readers have seen at the heart of Arthur's Order of the Round Table, and this is not surprising since nineteenth-century British culture was saturated with images of medieval chivalry, images that were "absorbed into the patterns of everyday life."[12] The idea of an Arthur with the qualities of a Victorian gentleman is one that closely connects the *Idylls* with popular culture.[13] Clearly, Tennyson in his Arthurian project was part of a broad cultural movement whereby idealized versions of "chivalry" were associated with attempts to improve society, especially in terms of reforming English masculinity. At the same time, the "barbaric" nature of medieval and ancient civilizations could not be ignored, and the ambivalence toward epic poetry studied by Dentith and others is part of this pattern. What I want to point out here is that Tennyson, like other artists and intellectuals of his day, was in this context grappling with universal aspects of human nature, and, as discussed in Chapter 1, the displacement of pre-state "cultures of honor," beginning in medieval Europe, had in fact led to rapid declines in homicide rates. For Tennyson and others, appeals to ancient traditions sometimes can be interpreted as innovative attempts to deal with ancient problems. Furthermore, the ambiguous relationships between moral standards associated with group identity – family, tribal, national, imperial – and Christian idealism can be enormously complicated. This helps to explain Tennyson's own conflicted attitudes toward the central figure of King Arthur, and I discuss this important issue in more detail below.

As John Ruskin observes in *Unto this Last*, the source of honor for the traditional soldier or warrior ultimately is self-sacrifice, not his willingness and ability to kill the enemy: "the soldier's trade, verily and essentially, is not slaying, but being slain" (VII: 36–7). Appropriately enough, Tennyson's Geraint, eventually restored to a healthy appreciation for his blameless wife, finally "crowned / A happy life with a fair death, and fell / Against the heathen of the Northern Sea / In battle" ("Geraint and Enid," 966–9).

Arthur himself, far from being a "female king," is a warrior–king, a powerful leader who, in spite of his gentle manner, has an enormous capacity for violence. In "Lancelot and Elaine," Lancelot describes Arthur at the Battle of Mount Baden:

---

[12] See Girouard, 146.

[13] Among many obvious and well-known examples of the association of specifically Arthurian "moral qualities" with conventional Victorian values in period art are the frescoes created by William Dyce for the Queen's Robing Room during the period 1847–64. See Girouard, 181.

> I myself beheld the King
> Charge at the head of all his Table Round,
> And all his legions crying Christ and him,
> And break them; and I saw him, after, stand
> High on a heap of slain, from spur to plume
> Red as the rising sun with heathen blood. (302–7)

Even after the final internecine bloodbath, a dying Arthur, surrounded by the corpses of his knights, when Bedivere is reluctant to follow his order to throw Excalibur into the lake, directs the following outburst to his last surviving knight:

> Unknightly, traitor-hearted! Woe is me!
> Authority forgets a dying king,
> Laid widow'd of the power in his eye
> That bow'd the will ...
> . . . .
> But, if thou spare to fling Excalibur,
> I will arise and slay thee with my hands.
> ("The Passing of Arthur," 288–300)

Arthur's relationship with his sword is the most obvious example of the fetishism concerning swords, armor, and "arms" shown by the knights throughout the *Idylls*. A great deal of Malory's warrior remains in Tennyson's Arthur – and yet readers complained about his "feminine" qualities. First, there is the implicit assumption that, in spite of his strength and courage, by his failure to challenge (and kill) Lancelot, Arthur has relinquished his manhood; this idea is, appropriately enough, articulated most succinctly by the cynical Vivien: "Man! is he man at all, who knows and winks? / Sees what his fair bride is and does, and winks?" ("Merlin and Vivien," 779–80). The issue of whether Arthur really *knows* remains unresolved in the poem, but it seems remarkable that he could not. However, there are broader problems for Tennyson in fusing the qualities of a mythic hero with those of a Victorian gentleman. His assumptions about manhood as a repudiation of "natural bestiality" preclude his adoption of a model that would incorporate anything like Charles Kingsley's positive "animal spirits" that inform "both marital vigor and sexual potency" in men.[14] Swinburne, with his remark about "morte d'Albert," was of course making a comment not only about Tennyson's obvious intention to associate Arthur with the Prince Consort but also about the values clustered around the ideal of bourgeois respectability,[15] an ideal that had gained increased status by contemporary associations with the monarchy. In the "Dedication," Tennyson refers to Albert as "modest, kindly, all-accomplish'd, wise, / With what sublime repression of himself" (17). These qualities are consistent with the dominant image of the Victorian gentleman as well as his predecessor,

---

[14] See James Eli Adams, *Dandies and Desert Saints: Styles of Victorian Masculinities*, 108.

[15] This is a key term in John Tosh's *A Man's Place*, discussed in Chapter 1.

the chivalrous knight. (In a famous portrait Albert himself was depicted in armor.)[16] Girouard divides midcentury attitudes toward the immensely popular Arthurian story into two groups, romantic and moral: "The romantics read Malory and were deeply moved and excited by the vividness of his stories of love, quests, fighting and marvels. The moralists saw Arthur and his knights as epitomizing (at their best) virtues which were still valid as a source of moral lessons for contemporary life" (180). In the *Idylls*, with the exception of the early romantic fantasy of "Gareth and Lynette," Tennyson emphasized the moral approach to his subject (a fact that by itself helps to explain the unsympathetic response of Swinburne and other contemporary critics with an aestheticist bias), and for Tennyson the most fundamental problem to be dealt with was the bestial nature of men. With his emphasis on the *repression* of spontaneous maleness (while preserving the male capacity for violence, appropriately controlled and directed), it is no wonder that Tennyson created a restrained and thoughtful Arthur who lacked the élan of Malory's hero. By placing Tennyson in the context of Victorian authors' efforts to "control the violence" of heroic characters, of course I do not mean to imply that this issue originated with them; on the contrary, they are dealing in new ways with an ancient problem of human nature relevant to Odysseus and Aeneas, for example.

Sussman says of Carlyle and his works: "In seeking a psychic armor to contain the inchoate, fluid energy within, [he] presents a particularly fragile and unstable model of the male psyche always at the edge of eruption, of dissolution, of madness" (19). Because Carlyle's answer to the problem of the unstable male psyche – monastic asceticism – is so far removed from Tennyson's, it may be surprising to find that Tennyson's formulation of the problem is very much the same (although, one could fairly argue, more extreme than Carlyle's). Nearly all the knights (finally even Lancelot) exhibit at least the potential for explosive, irrational violence. Arthur's "perfection" is most evident in his ability to control his own considerable capacity for violence.

However, the relationship between madness and unregulated male violence is examined most closely (even schematically) in "Balin and Balan," the fifth idyll in the final order but written last. In this little-discussed story, Balin, called "the Savage," comes to Camelot in the company of his older and more sophisticated brother, Balan, seeking to acquire courtesy as a cure for his frequent violent moods. Balin worships the beautiful and supposedly virtuous queen and strives to emulate the courteous behavior of the greatest knight. Their example helps him to hold his violent nature in check. However, while his brother is gone on a quest to rid a forest of the demon that haunts it, Balin inadvertently observes a compromising scene between Lancelot and Guinevere. The shock to Balin is parallel to that of Pelleas in the fifth idyll when he discovers Ettarre and Gawain asleep together in sin, the shock that triggers his descent into total disillusionment and his metamorphosis into the Red Knight. Without the steadying influence

---

[16] On chivalry and the royal family, see Girouard, 112–28.

of Balan, this traumatic incident, coupled with the scandalous rumors spread by Vivien, leads to a recurrence of Balin's madness. In a violent rage, he rides into the forest, where he hopes to destroy the demon, reasoning that "To lay that devil would lay the Devil in me" (296). Balin comes upon King Pellam's castle, where he is taunted by the knights, who have heard of the scandal at Camelot, for displaying on his shield the crest of "The Queen we worship, Lancelot, I, and all, / As fairest, best and purest" (344–5). In particular, Sir Garlon mocks the "fair wife-worship" that "cloaks a secret shame" (355). Losing his precarious self-control, Balin – not in a "fair fight" but in a sudden, spontaneous move – "Hard upon helm smote him ... / Then Garlon, reeling slowly backward, fell, / And Balin by the banneret of his helm / Dragged him, and struck" (390–3). Because Garlon has been vilified in Tennyson's description (objectified, dehumanized, much like the Earl of Doorm in Geraint's story), Balin, despite his uncontrolled, murderous rage, can be seen as a sympathetic figure as he flees the unchivalrous castle of Pellam. Escaping into the forest, where he discards his shield with Guinevere's crest, Balin has fallen into bestiality, like Dr. Jekyll receding into Mr. Hyde. He finally meets his brother, who mistakes him for the demon and attacks. The resulting double fratricide foreshadows the tale of the ancient unnamed king and his brother that is embedded in "Lancelot and Elaine." In Freudian terms, this is a symbolic suicide: Balan and Balin have been consistently portrayed as a divided self, and in the end the rational self (Balan) can control the savage, bestial self (Balin) only by destroying it (Rosenberg, 82).

As Rosenberg points out, "two kinds of time ... cyclic and apocalyptic ... are built into the structure of the *Idylls* (37). Arthur himself is associated with apocalyptic time, but the "new order" he introduces by conquering the heathen, driving out the beast, and felling the forest, as described in "The Coming of Arthur" (58–62), is itself displaced by an even more barbarous one after the "last dim battle in the West." Arthur's great plan is graphically laid out in the symbols incorporated into Merlin's "four great zones of sculpture" girding the great hall at Camelot:

> And in the lowest beasts are slaying men,
> And in the second men are slaying beasts,
> And on the third are warriors, perfect men,
> And on the fourth are men with growing wings
> ("The Holy Grail," 234–7)

It is as though in establishing his order, instead of building on previous orders, Arthur must begin from the beginning, and Tennyson appeals to the primal human fear of beasts that recalls a prehistoric time before mankind dominated the earth, when human beings were hunted prey as well as hunters. In this sense Tennyson implies that Arthur's order corresponds to an entire cycle of civilization.

This may seem surprising because Tennyson also encourages his readers to think of Arthur's historical displacement of the Roman invaders and his defeat of the (Saxon) pagans. However, it is important for Tennyson to associate his mythic

Arthur with the ur-story of civilization because it is symbolic of the continuing internal struggle of *men* with *the beast within*, Tennyson's principal focus in the narrative, inextricably tied to his theme of morality. The fundamental problem for Arthur's Order of the Round Table is to provide the ritual and teach and enforce the discipline needed to control the (natural) male bestiality and madness that Tennyson, in reaction to the legacy of his Romantic predecessors, associated with savage violence and moral anarchy. Tennyson focuses on irrational, instinctual male violence as the primary source of civilization's ills.

In the *Idylls* Arthur's origin is shrouded in mystery – supposedly he is the son of Uther (and therefore the legitimate heir to his throne), but doubt remains. If he *is* Uther's son, then he is literally the product of savage male violence: Uther went to war with the "prince and warrior" Gorloïs because he lusted after Gorloïs's wife, Ygerne, and she spurned his advances. After killing Gorloïs in battle, "Uther in his wrath and heat besieged / Ygerne within [the castle] Tintagil" and, after her outnumbered defenders deserted her, forced her "to wed him in her tears, / And with a shameful swiftness" ("The Coming of Arthur," 198–204). If King Uther is not a murderer and rapist, he is something very close to that, and this fact, along with Tennyson's scheme of myth-making, helps to explain why Tennyson leaves the origins of Arthur, the apotheosis of honorable manhood, shrouded in mystery. Arthur's transcendence of ordinary maleness (even as manifested in a savage king) in his supposedly perfect, Christ-like manhood is further emphasized. The most coherent explanation of Arthur's origin is given by Bedivere, the first and last of Arthur's knights, to King Leodogran, Guinevere's father: Arthur, son of Uther, was born "before his time" to Ygerne and immediately "Deliver'd at a secret postern-gate / To Merlin" (210–12). (And yet both Uther and Ygerne are dark-complexioned, whereas Arthur is strikingly blond.) According to Bedivere, his account is challenged by those who call him "baseborn, and since his ways are sweet, / And theirs are bestial, hold him less than man: / And there be those who deem him more than man, / And dream he dropt from heaven" (179–82). However, the only sure basis for Arthur's kingship is his charismatic, manly assertion of moral authority – binding his men by "strait vows to his own self" (261) and his ability to govern effectively.

Also, it is easy to speculate about underlying personal issues in Tennyson's handling of both male madness and the challenge to patrilinal authority that Gilbert believes makes Arthur a "female." Dr. George Clayton Tennyson, Tennyson's clergyman father (and boyhood tutor) was virtually disinherited by his own father (of the same name) in favor of the younger son Charles – a family tragedy that was always associated with Dr. Tennyson's subsequent alcoholism and suicidal madness. In 1820, when Alfred was 10, "relations between his father and grandfather became morbidly irreparable" and in the 1820s Dr. Tennyson "collapsed into drink and rage" (Ricks 4). And "[w]hether their father was fully responsible for it or not, the minds of most of the … children were not to prove stable" (Ricks, 57).

In *Maud* (1855), Tennyson's most explicit exploration of male madness (and his favorite poem for reading aloud), irrational and destructive male violence is closely associated with male sexuality, and the ending (disturbing to many modern readers and critics) suggests that the unstable protagonist (guilty of killing his lover's brother in a duel) will redeem himself by immersion in the righteous and patriotic cause of the Crimean War. Even to contemporary readers, "[t]he speaker's bellicosity and actual fighting seemed ... hysterical and the war to reinstate 'The glory of manhood' ... seemed out of proportion to its grounding in the speaker's experience" (Sinfield 176). Early in the narrative, the protagonist is increasingly smitten by Maud as he overhears her singing "A passionate ballad ... / A martial song ... / Singing of men that in battle array / ... / March with banner and bugle and fife / To the death, for their native land" (165–72).

In the *Idylls,* potentially destructive male energy is curbed and controlled by the rigid chivalric code instituted by Arthur, and its primary regulating instrument is woman-worship, a code that assumes the moral superiority of women and places a premium on female life and welfare. The chivalry of the Round Table – which must be continually reaffirmed both through the ritualized court violence of the tournament and the knightly quest into the world in order to rescue women and redress their wrongs – is assumed to be consistent with the primitive, unarticulated imperative of protecting women (because of their childbearing and child-rearing functions) as the basis of tribal or national survival and also, on a more abstract, idealized, and articulated level, as the basis of civilization and its triumph over barbarism. But if the *Idylls* are to incorporate a serious moral vision for his time, Tennyson's version of chivalry must also conform to standards of bourgeois respectability, in particular the domestic, moral virtues of womanhood that we now routinely associate with the title of Coventry Patmore's poem "The Angel of the House"[17] and so fully articulated by Ruskin in his essay "Of Queens' Gardens" ("It is the type of an eternal truth – that the soul's armor is never well set to the heart unless a woman's hand has braced it; and it is only when she braces it loosely that the honour of manhood fails") (120). Gender roles in the *Idylls* are in fact largely consistent with these stereotypes, whereby wives, through their innate moral superiority and freedom from the contamination of the world, have the power to ennoble their husbands, ameliorating the corrosive effects of male insensitivity and male aggression. It is obvious from the beginning of the *Idylls* that Tennyson rejects the ideal of "monastic asceticism" found by Sussman and J. E. Adams in the works of Carlyle, Pater, and the Pre-Raphaelite Brotherhood: after being smitten by the light of Guinevere's eyes, Arthur believes that:

> '... saving I be join'd
> To her that is the fairest under heaven,
> I seem as nothing in the mighty world,
> And cannot will my will, nor work by work
> Wholly, nor make myself in mine own realm
> Victor and lord. But were I join'd with her,

---

[17] For a representative feminist treatment of this issue, see Carol Christ's well-known essay "Victorian Masculinity and the Angel in the House."

> Then might we live together as one life,
> And reigning with one will in everything
> Have power on this dark land to lighten it,
> And power on this dead world to make it live.'
> ("The Coming of Arthur," 85–93)

Later, in "Balin and Balan," the all-male court of King Pellam (who in his misguided desire for piety has banished all women including his own "faithful wife") is described as sterile and gloomy. And yet Arthur's idealized vision of monarchy and matrimony is doomed from the beginning. His instantaneous love for Guinevere is not reciprocated. Arthur, modestly dressed and without helm or shield, does not stand out in the parade of warriors before Leodogran's castle, and "She saw him not, or mark'd not, if she saw" ("The Coming of Arthur," 53). Later, she falls in love with the King's lieutenant, the sexually attractive Lancelot, as he escorts her to Arthur's court. The passive and even unintentional conquest of Arthur by the beautiful Guinevere is parallel to the equally effortless conquest of the naive and idealistic Pelleas by the utterly self-centered and manipulative Ettarre in the ninth idyll. Just prior to his first sight of Ettarre, Pelleas is fantasizing about his ideal love, and, ignorant of the Queen's adultery, he imagines his lady to be "fair ... and pure as Guinevere, / And I will make thee with my spear and sword / As famous – O my Queen, my Guinevere, / For I will be thine Arthur when we meet'" ("Pelleas and Ettarre," 42–5). When Ettarre and her entourage suddenly appear, "Pelleas gazing thought, / Is Guinevere herself so beautiful?" (64–5). Clearly, this incident is a retelling of the first meeting of the King and Queen, just as the idyll as a whole is a retelling of "Gareth and Lynette," the second idyll and the only one with the thorough-going fantastical quality of a fairytale (in which the persistence of the dedicated knightly lover pays off and the violence of combat turns out to be imaginary rather than real). But now it is feminine goodness and knightly honor that are illusory, and after being betrayed by both his lady and his fellow knight Gawain (the two of whom, he discovers, have been engaging in casual sex) and then learning of the Guinevere–Lancelot liaison, the disillusioned idealist descends into savage madness. In the next idyll, "The Last Tournament," Pelleas re-emerges as the Red Knight, who, cynically inverting the professed values of the Round Table, mocks Arthur as a "woman-worshipper," a "Eunuch-hearted king / Who fain had clipt free manhood from the world" ("The Last Tournament," 444–5). His Round Table in the North is made up of adulterous knights and their ladies are all harlots, but since they all openly admit the truth, the embittered idealist has created a space in which professed moral values conform to reality.

Tennyson's use of the Red Knight as a kind of anti-Arthur illustrates his surprising distance from the melodramatic imagination central to much of Victorian literature, and he shows a great deal of moral ambivalence here. Of course, Arthur faces his nemesis with predictable, intrepid courage, but the drunken Red Knight practically destroys himself, toppling over into the mud after an inept charge at the King, and the royal knights immediately leap on his fallen body. They "trampled out his face from being known, / And sank his head in mire, and slimed themselves" ("The Last Tournament," 467–70) before going on to massacre the entire party of rebellious knights and apparently even the women as well. Symbolically, the ostensible villain has won because the Knights of the Round Table in fact fight like wild beasts and, in spite of their professed code of honor, reveal themselves to be no better than uncivilized madmen. Most important is Arthur's consciousness of his failure: just before the fighting began, Arthur had begun to recognize the Red Knight as the once-noble Pelleas, and after the bloodbath, we are told that "all the ways were safe from shore to shore, / But in the heart of Arthur pain was lord" (485–6), a passage that parodies the description of Arthur's early triumph (also bloody but in this case, apparently glorious) over the petty feuding barons and kings in order to consolidate his power: "And in the heart of Arthur joy was lord" ("The Coming of Arthur," 123).

While Arthur's men were disgracing themselves in battle, the parallel ritualistic combat of the "last tournament," presided over by a demoralized Lancelot, who passively observes "the laws that ruled the tournament / Broken" (160–61) and won by the cynical Tristram, is itself a pathetic parody of bygone glory. Moral anarchy has returned, the Order of the Round Table is effectively dead, and Tennyson sees no need to portray the events that led up to the actual civil war instigated by Modred.

In the penultimate idyll, a reproachful but forgiving (and still loving) Arthur, just prior to the final, cataclysmic battle, confronts a penitent Guinevere in the convent where she has retreated. William Morris and generations of critics have defended Guinevere against "priggish" Arthur's charge that she personally is to blame for the destruction of the Round Table – "For thou hast spoilt the purpose of my life" (450) – but Tennyson's narrative has shown this to be quite true. In Arthur's system, women wield enormous power: through their innately superior morality ("purity," goodness) – as symbolized by their physical beauty – women have the social function of regulating, controlling the potentially bestial, innate sexual energy of men.

As Nancy F. Cott has argued, between the seventeenth and the nineteenth centuries, the "dominant Anglo-American definition of women as *especially* sexual" was reversed and transformed into the view that women "were *less* carnal and lustful than men" (221). In particular, Evangelical Protestantism increasingly emphasized the idea that Christianity had elevated women above the sensuality of animal nature and made them "more pure than men" (227). Gilbert uses this idea of the apparently unnatural suppression of female sexuality to explain the destruction of Arthur's order through an outbreak of the "irrational and destructive" natural forces associated with female sexuality. Although it is obvious that the suppression of female sexuality is central to Arthur's woman-worship, Gilbert somehow misses the crucial point that femininity is blameworthy in the *Idylls* when it fails to control the innate, irrational, and destructive forces of the *male*. It is true that Guinevere is sexual, that she craves the love of man who (unlike Arthur) has "a touch of earth" ("Lancelot and Elaine," 133), and she is also self-centered and small-minded: Arthur is "A moral child without the craft to rule, / Else had he not lost me" ("Lancelot and Elaine," 145–6). In managing her adulterous affair with Lancelot, she manipulates the gullible and trusting Arthur, but she is by no means a demon lover, nor does she possess extraordinary or mysterious powers beyond her charismatic physical beauty. She is a *femme fatale* precisely because she fails to conform to the idealized womanhood projected upon her by Arthur and the idealistic knights who worship her as a model of feminine beauty and (therefore) goodness. Although Tennyson portrays her somewhat sympathetically as having the grace and refinement befitting a queen, her feminine "powers" do not differ in kind from those of the shameless Ettarre.

Even Vivien is not portrayed as a demon but rather as a cynical manipulator of men and perhaps a psychopath.[18] And although she is depicted as a "harlot" who is more than willing to use sex to attain her aims (even making a futile attempt to seduce Arthur himself), it is misleading to term her victimization of Merlin as a sexual seduction (or the victory of feminine sexuality over masculine wisdom) as critics routinely do.[19] Her persistent flirting, crying, and wheedling, intensified by wounded pride in response to his calling her a harlot to her face, finally exhausts Merlin, who in any case has already accepted the inevitability of the final battle in the west, and, "overtalk'd and overworn," he capitulates and reveals the secret charm, which he expects will be used to entrap him. Although gender roles are operative here, Merlin's failure to protect himself has little or nothing to do with his sexual lust. Vivien's "power" rests in her unalloyed ruthlessness and skill in manipulating Merlin so that he succumbs to his own magic.

If Arthur's judgment of Guinevere seems harsh, it may be because the reader senses that a dysfunctional ideology rather than personal failure is the root problem in Camelot. In a sense, Guinevere, Tristram, and others who complain about the impossibly strict vows required by Arthur are right. "[C]an Arthur make me pure / As any maiden child?" asks Tristram ("The Last Tournament," 687–8), who reasons that in the beginning the vows, "the wholesome madness of an hour" (an interesting choice of words in the context of male madness) inspired each knight to reach beyond himself doing:

> mightier deeds than elsewise he had done,
> And so the realm was made; but then their vows –
> First mainly thro' that sullying of our Queen –
> Began to gall the knighthood (670–78)

---

[18] Tennyson uses some demonic imagery in describing an angry Vivien (ready to stab Merlin if she had had a dagger), as when "the bare-grinning skeleton of death" flashed from her "rosy lips of love" (844–5). Nevertheless, the tools Vivien successfully uses to "entrap" her victim are those of psychological manipulation, realistically portrayed, and not directly related to sexual seduction.

[19] For example, see Ryals's discussion of the "seduction scene." Ryals is correct in commenting that "here the veil of Victorian reticence about sexual matters was lifted for the moment" (42), and in fact Tennyson's contemporaries were quick to notice the uncharacteristic naughtiness in Tennyson's treatment of Vivien's sexuality. Nevertheless, Merlin does not yield to Vivien's request out of lust.

and so began the downhill slide. As Clyde de L. Ryals puts it, "At the end Arthur learns that the world is impregnable to morality" (89), and "Arthur does not redeem the world because the world is irredeemable" (92). This view is consistent with the traditional interpretation of the *Idylls* as the narrative of the collapse of a society that rejects spiritual values, but it assumes that the collapse is inevitable, and indeed there is evidence even in the first idyll to suggest that this is the case.[20] Even Gilbert asserts that the poem "is certainly about the decline of a community from an original ideal state" and goes on to locate "human sexuality and, in particular, female passion" as "an important agency in this decline" (864).

In my view, however, Arthur's formulation of "spiritual ideals" themselves is confused and unstable from the beginning, and, although Tennyson obviously prized spirituality and believed that the imperfections of human nature hindered mankind from achieving it, he also intuited structural problems in the dominant social ideologies of his day which were supposed to support spiritual values. In the dreamlike, mythic *Idylls,* Tennyson could have it both ways, for, as Freud pointed out, dreams have a strange tendency to "disregard negation and to express contraries by identical means of representation" (44–5). In the list of knightly oaths given in "Guinevere," Arthur in effect summarizes his "high ideals":

> To reverence the King, as if he were
> Their conscience, and their conscience as their King,
> To break the heathen and uphold the Christ,
> To ride abroad redressing human wrongs.
> To speak no slander, no, nor listen to it,
> To honour his own word as if his God's,
> To lead sweet lives in purest chastity,
> To love one maiden only, cleave to her,
> And worship her by years of noble deeds,
> Until they won her; for indeed I knew
> Of no more subtle master under heaven
> Than is the maiden passion for a maid,
> Not only to keep down the base in man,
> But teach high thought, and amiable words
> And courtliness, and the desire of fame,
> And love of truth, and all that makes a man. (465–80)

Aside from illustrating Arthur's flaw of seeing his knights "as but the projection of himself" (Ryals, 89) – which in spite of Arthur's "sweetness" would suggest a tyrant's cult of personality – this statement is very problematical in light of the narrative of "moral decline" that will soon reach its climax in Arthur's profound self-doubts as the Round Table is finally annihilated.

---

[20] Rosenberg writes, "From [a] cyclic perspective, man's reeling back into the beast is both monstrous and *natural*. The Round Table is founded to arrest this process, and it succeeds only 'for a space'" (37).

King-worship and woman-worship are inadequate and even destructive ideals. When young Pelleas sees Ettarre, "The beauty of her flesh abash'd the boy, / As tho' it were her soul" (74–5). Physical beauty does not correspond to spiritual goodness, which is essentially asexual, as seen in the extreme examples of Percival's sister, celibate in her convent, and Galahad, apparently translated into the New Jerusalem in "The Holy Grail." Less extreme is the example set by Percival, who appropriately chooses the life of Ambrosius and the other good, holy men in a celibate male order that necessarily remains marginal to the larger society (as Arthur well understands). In the world of the *Idylls*, in spite of Arthur's ideology, women are not innately morally superior to men. However, they are not monsters, either: if they wield enormous destructive powers, it is because of the false images that are projected on them by the King and his knights. Even the basic value of truth-telling is immediately compromised when the "truth" about Guinevere becomes a lie and "slander" becomes the truth. As for "fame," its hollowness is emphasized as Merlin finally cedes his own to Vivien.[21]

The 12 great battles in which Arthur and his men defeated the "Godless hosts" ("heathen," "pagans") are not represented in Tennyson's narrative except in retrospective reporting. At the beginning of the narrative proper, they are already in the "glorious past," and the *Idylls* are rather concerned with the maintenance of (or rather the failure to maintain) the Order of the Round Table. One very significant implication of the *Idylls* narrative is that *ideals are sustainable only during a time of war.* We are reminded of Hallam Tennyson's comment that "My father felt strongly that only under the inspiration of ideals, and with his 'sword bathed in heaven,' can a man combat the cynical indifference, the intellectual selfishness, the sloth of will, the utilitarian materialism of a transition age" (II:129). Even the apparently glorious legacy of righteous slaughter underwritten by Arthur's unquestioned idealism is not without its darker side, however, as suggested by its analogue in the massacre of the Red Knight and his followers and by the insidious function of the orphan Vivien, who tells Guinevere, "My father died in battle for thy King, / My mother on his corpse – in open field" ("Merlin and Vivien," 70–71). Arthur assumes that all of his professed values derive from or are at least consistent with Christianity and that his purpose writ large is to "uphold the Christ," that is, establish a Christian civilization, but it is unclear that either the worship of the King or woman-worship is entirely consistent with this assumption. At any rate, the operative principle of love in Camelot is eros rather than agape. Moreover, adultery had always been part of the heritage of chivalry.[22] In part, Tennyson's Arthur represents the masculine fear that masculine purity and goodness are not

---

[21] It is clear that Vivien's principal motive in destroying Merlin is to steal his fame or glory: "I have made his glory 'mine,' / . . . 'O fool!'" she shrieks (969–70).

[22] Extramarital game-playing among the upper classes in Victorian times was often associated with chivalric images and themes. See Girouard (178–218). As Girouard points out, "[p]urity as an element of modern chivalry," so important to Tennyson, "only began to loom large in the 1850s, when it was grafted onto earlier concepts of chivalrous love, which included "devotion, tenderness, courtesy and protection" (199). Kenelm Digby's

sexually attractive to women. But Tennyson also implies that even the greatest physical prowess and courage (displayed by Arthur on the battlefield and in his *useful* work as King) are inadequate unless *performed* (as in the tournament, where Lancelot excels). And yet if this generalization holds it would run counter to the values of bourgeois respectability. Seen in the context of Tennyson's structural problems of artistic representation, the chivalric plot of masculine initiation and quest does not mesh well with the marriage plot (associated with bourgeois respectability) that dominates Victorian fiction.[23]

In the final analysis, however, it is the failure of Arthur's order (based on his flawed ideal of manhood) to control the irrational force of male energy and the annihilating effects of male violence that leads to his downfall. As he marches with his men to the "last weird battle in the west" and the pathetic end of his reign, Arthur moans that God has forsaken him:

> "I found Him in the shining of the stars,
> I marked Him in the flowering of His fields,
> But in His ways with men I find Him not.
> I waged His wars, and now I pass and die."
> ("The Passing of Arthur," 9–12)

To the disoriented and confused Arthur, it seems as though "some lesser god had made the world, / But had not force to shape it as he would, / Till the High God behold it from beyond, / And enter it, and make it beautiful" (14–7). The "last, dim, weird battle of the west" (94) itself is described in terms that remind us of the battle of Armageddon in *Revelation* as well as Matthew Arnold's "darkling plain / Swept with confused alarms of struggle and flight / Where ignorant armies clash by night" (35–7) in "Dover Beach":[24]

> A deathwhite mist slept over sand and sea:
> Whereof the chill, to him who breathed it, drew
> Down with his blood, till all his heart was cold
> With formless fear; and even on Arthur fell
> Confusion, since he saw not whom he fought.
> For friend slew friend not knowing whom he slew;
> ....

The Broad Stone of Honour (1822), enormously influential in popularizing chivalry among young men, has little to say about purity.

[23] Another problem for Tennyson in adapting chivalry to his purpose is the elitist tendency in chivalric traditions, but this is part of a very large and complex pattern of cultural assimilation that involves all Victorian uses of chivalry and the ideal of the gentleman. According to J. E. Adams, "[t]he gentleman ... is the most pivotal and contested norm of mid-Victorian masculinity because it served so effectively as a means of regulating social mobility and its attendant privileges" (152).

[24] Tennyson's "deathwhite mist" is an addition to Malory. Arnold's image of ignorant armies clashing by night is the best-known Victorian reference to the Battle of Epipolae as described by Thucydides.

> And ever and anon with host to host
> Shocks, and the splintering spear, the hard mail hewn,
> Shield-breakings, and the clash or brands, the crash
> Of battleaxes on shattered helms, and shrieks
> After the Christ, of those who falling down
> Looked up for heaven, and only saw the mist;
> And shouts of heathen and the traitor knights,
> Oaths, insult, filth, and monstrous blasphemies,
> Sweat, writhings, anguish, labouring of the lungs
> In that close mist, and cryings for the light,
> Moans of the dying, and voices of the dead. (95–117)

Before concluding the slaughter by killing Modred, Arthur tells Bedivere:

> ... on my heart hath fallen
> Confusion, till I know not what I am,
> Nor whence I am, nor whether I be a king;
> Behold, I seem but king among the dead. (143–6)

These passages, from the 1869 additions to the original 1842 version of "The Passing of Arthur," reinforce the impression that the *Idylls* can be read as an exploration of dilemmas, uncertainties, contradictions.

Following what he understood to be his father's moral scheme in the *Idylls* (whereby the sin of Guinevere "spreads its poison through the whole community"), Hallam Tennyson referred to Mark's murder of Tristram (struck from the rear, Tristram is "cloven through the brain") after finding the knight in a compromising position with his wife Isolt ("The Last Tournament," 747–8):

> We have here the deadly proof of the kinship of all wilful sin in murder following adultery in closest relation of cause and consequence,--the prelude of the final act of the tragedy which culminates in the temporary triumph of evil, the confusion of moral order, closing in the great "Battle of the West." (II:131)

If taken seriously, the vaguely biblical idea that the sin of adultery can be expiated only by the shedding of blood is particularly troubling because of its implications for the Christian doctrine of the Atonement. It suggests that Arthur's order may be pre-Christian rather than Christian and that the doubts about his Godly mission expressed by Arthur toward the end may be well founded. But no resolution is offered and moral confusion is not resolved but rather evaded by the transition into the formulaic mythic narrative of the ending, one in which mystery and irresolution is not only acceptable but essential, and Arthur returns to Avalon or "the great deep" or wherever.

I am convinced that Tennyson's doubts not only about social constructions of gender but about the foundations of Western civilization are increasingly reflected in his later revisions to his enormously suggestive masterwork. However, in offering this reading of the *Idylls*, I am not arguing that it was necessarily the primary one intended by Tennyson, who continued to associate the "new-old" legend of Arthur with the Ideal and with meliorist hopes for future spiritual progress. Instead, I appeal to T. S. Eliot's insight that Tennyson, "the saddest of all English poets," was "the most instinctive rebel against the society in which he was the most perfect conformist" (*Essays*, 189).[25] And on the most basic level, I want to show how Tennyson was working with the predispositions associated with human nature, as discussed in Chapter 1.

In the 1960s and 70s, a reaction against the modernist deprecation of Victorian literature and culture associated with the work of Houghton, Buckley, and the others, also led to sympathetic reevaluations of the *Idylls*, which made very large claims for what can be seen as the masterwork of the most important Victorian poet. For example, Ryals defended Tennyson's blend of dramatic, narrative, and lyric genres in the poem and even argued that Tennyson had "solved, to a great degree, the problem of the long poem for modern literature" (54). On a thematic level, Ryals asserted that "[i]n the *Idylls* Tennyson ... takes into account all the contradictions of existence. For the poem ... finally shows life as a permanent tension between the finite and the infinite" (200). I am sympathetic with Ryals's continuation of the (Victorian) humanistic search for universal truths in Tennyson, but surely it is extravagant to claim that Tennyson takes all the contradictions of existence into account. Writing about a decade later, Rosenberg made, it seems to me, more tenable claims for Tennyson as an important forerunner of the Symbolist Movement, though his claims for the "remarkable unity" of the poem (13) may be overstated. For him the *Idylls* is "one of the four or five indisputably great poems

---

[25] As I have implied earlier in this chapter, Eliot was not especially perceptive about the *Idylls*; however, his insight concerning the apparently conformist Tennyson's instinctive rebelliousness helped to inspire studies like E. D. H. Johnson's *The Alien Vision of Victorian Poetry*, which argued that Tennyson had attempted to reconcile his need to maintain an aesthetic integrity with his need to satisfy a public whose demands were often at odds with his artistic vision. Terry Eagleton's more recent, politicized version of Tennyson's doubleness is that "Tennyson is ... at once a poet of the 'centre' and the 'margins': on the one hand, poet laureate, spokesman for conservative values and Victorian patriarchy; on the other hand, a radically alienated, deeply subjective refugee from the march of bourgeois progress, whose "feminine" sensuousness finds its nourishment in the privatized and exotic" ("Editor's Preface," Sinfield, ix). More recently, Catherine Phillips observes, "The polyphony of the *Idylls* allows for participation in contemporary debate: the poem displays the chivalric code and Christian values that the decoration of Parliament vaunted but also contains a profound, contemporary understanding of human psychology" (252). Of course, in my analysis, I am not concerned primarily with the ambivalences of an instinctively rebellious and aesthetic young Tennyson but rather with a mature sage figure brooding about the past and future of England and of humankind. Eliot writes, "I do not believe for a moment that Tennyson was a man of mild feelings or weak passions. There is no evidence in his poetry that he knew the experience of violent passion for a woman: but there is plenty of evidence of emotional intensity and violence – but of emotion so deeply suppressed, even from himself, as to tend rather towards the blackest melancholia than towards dramatic action" (181). I would argue that Tennyson's melancholia is evident in both the narrative form and the expressive language of the *Idylls*.

in our language," but he describes it as "Tennyson's doom-laden prophecy of the fall of the West" (1). Because his insights are especially pertinent to the line of analysis I have pursued in this study, I will quote Rosenberg on this prophetic power of the poem at some length. After discussing T. S. Eliot's dismissal of the *Idylls* ("Tennyson could not tell a story at all" [5]), Rosenberg continues:

> Although modern literature offers no more chilling prospect than the closing books of the *Idylls*, some of the finest minds of the past generation have rejected the poem as mawkishly insincere. Blindness from such sources can itself cast light, as in the case of Harold Nicolson's attempt to cope with the *Idylls* in 1923, an effort less disingenuous than Eliot's a decade later. Nicolson admits to a strong personal admiration for the poem but reserves final judgment on the grounds that, despite the "magnificent poetry," its effect can only "be one of estrangement and hostility, since it is impossible for us to conquer the impression (doubtless an incorrect and transitory impression) that these poems ... are for the most part intellectually insincere." And he goes on to suggest that, in view of "the peculiar adjustment to which our nervous system has attained in this year 1923," the *Idylls* be the first of Tennyson's works dropped from any collection seeking to interest the modern reader.

Rosenberg observes that Nicolson's dismissal of the *Idylls*:

> makes perfect sense for the nervous system of 1923 once one recognizes that the system was in a state of shock from the First World War – a catastrophe of which the *Idylls* is the profoundest sense a prophecy – and the seemingly rock solid values of the Victorians had proved as ephemeral as Camelot itself. (6)

I think that Rosenberg is very perceptive here in describing the poem as a prophecy of the First World War; however, his analysis of the *Idylls*, although full of insights about Tennyson's method in developing the apocalyptic implications of the "last, dim, weird battle of the west," does not explore the profound associations of violence with "manhood" at every level in the poem and the implications they have for the dismal failure of Arthur's Round Table as a social order. These associations are particularly telling in the larger cultural context where the ideals of chivalry have often been linked to the origins of the Great War.[26]

In Chapter 1, I outlined my assumptions concerning the relevance of human nature in discussing literary representations of human psychology and behavior and referred specifically to research that helps us to understand the cogency of

---

[26] As Girouard puts it, "Opinions will always differ as to whether the Great War could or should have been prevented. But one conclusion is undeniable: the ideas of chivalry worked with one accord in favour of war" (276). Michael C. C. Adams discusses similar ideas at some length in *The Great Adventure*: for example, "Chivalry obfuscated the inhuman quality of modern war and in so doing it not only failed to contain slaughter but helped to encourage it" (72). Paul Fussell's discussion of the disjunction between the naive, chivalric language of Tennyson's Arthurian poems and other pre-war romances and the realities of modern warfare in *The Great War and Modern Memory* is especially provocative, but I am arguing that Tennyson himself is a prophet of the disillusionment Fussell describes.

Tennyson's explorations of male violence as related to social order and sexuality. His fundamental emphasis on "driving out the beast" through a Christian religious discipline that somehow transcends nature, allied with the ideology of chivalry, which enforces monogamy and discipline among men and thus promotes social stability, can be described in terms of culture and social construction (although Christianity is associated with "universal" human concerns), and his anxiety concerning the implications of his narrative about a mythic order torn apart by its internal tensions and contradictions for Victorian Britain and its Empire is historically specific. However, any serious consideration of Tennyson's anxious, troubled representations of innate urges – including inclinations toward violence – should be informed by an awareness and recognition of what is known and theorized about human beings in the natural world. Earlier I cited a body of relevant scientific literature on topics such as male-male competition, dominance hierarchy, and mating and reproductive strategies,[27] but here I want to consider related issues in terms of Tennyson's literary imagination of human nature. If the legendary Arthur is seen as a "real man," his life-history is problematic indeed. Most obviously his mythic entry into and exit from the world are not realistic, and the resonance of this fantastical quality of his life-history is at odds with the very worldly issues such as marital fidelity at the heart of the story, issues that according to the mocking comments by Swinburne quoted at the beginning of this chapter, for example, suggest that Tennyson may as well have been writing a Victorian sensation novel or melodrama. It is true that the epic poem shares some characteristics with the genre of Victorian fiction, for example, crucial problems involving "theory of mind," as Arthur and Guinevere each fail to adequately assess the thoughts, feelings, and motives of the other. However, the serious intention that Swinburne missed is Tennyson's sincere engagement with the concept of an idealized male hero who dramatically "fails" as a man. It is not just the implicit shame of "cuckoldry" as a result of his wife's longstanding affair with his best friend and trusted lieutenant. (In fact, the positive "male bonding" between Arthur and Lancelot, in line with the knightly fellowship of the Round Table and in contrast to the failure of the royal marriage and the treacherous and disastrous sin of Lancelot and Guinevere, deserves further study.)[28] Arthur's failure to produce

---

[27] In addition to Daly and Wilson, cited in Chapter 1, see, for example, B. B. Smuts, D. L. Cheney, R. M. Seyfarth, R. W. Wrangham, and T. T. Struhsaker, eds., *Primate Societies* (1987); J. M. Plavcan, C. P. Schaik, and P. M. Kappler, "Competition, Coalitions and Canine Size in Primates" (1995); M. P. Ghiglieri, "Sociobiology of the Great Apes and the Hominid Ancestor" (1987). Contrary to a popular a misconception, insights into evolved biological differences between men and women do not inevitably lead to the conclusion that these differences are immutable. (See Geary, *Male, Female*, 330–31.)

[28] In his essay "Male Bonding in the Epics and Romances," Robin Fox applies the theory of male bonding, which originated with Lionel Tiger in his book *Men in Groups* (1968), to key male characters in several traditional epics, including Tennyson's primary source, Malory's *Morte d'Arthur*, where he finds the "deep emotional, even spiritual, bonds between the knights" as central to the story, while "romantic love interest is intrusive and destructive of this sacred bond" (138).

an heir is magnified by his role as monarch: the civil war initiated by his "evil nephew" Modred is fatal to the King and his Order, and this tragedy is consistent with the "personal" failure of the King's marriage and his failure to establish a dynasty. This failure is related to Arthur's uncertain origins and the confused kinship relations that characterize this narrative. A social order that apparently represents the zenith of human civilization disintegrates as a result, and it makes sense that Tennyson took care to disconnect Arthur's Order, however glorious in its mythical heyday, from that of Victorian Britain. However, Tennyson implies that Arthur's tragedy is meaningful, not just to the Victorians or to the British in particular, but to mankind.

At this point I will give an interpretive summary of the *Idylls* in Tennyson's final arrangement, emphasizing the behavioral systems discussed in Chapter 1 and showing how the individual narratives of the idyll episodes function within the overall framework of the master narrative of King Arthur and his Order. Initially, in *The Coming of Arthur*, Arthur woos and wins the hand of the princess Guinevere, daughter of King Leodogran, and in bringing home his queen to Camelot, reinforces his own status and sense of well-being in his new kingdom based on a uniquely coherent and law-abiding Order. According to Hogan's scheme, as previously discussed, the predominant prototypes for stories ending in happiness are "romantic union with one's beloved" and "the achievement of political and social power." In an obvious way, the initial idyll can be classified as both a romantic and heroic comedy. However, with Arthur's traitorous nephew Modred lurking in the background, there is potential for tragedy in the master narrative, and, as pointed out, in the long process of writing and publishing the individual idylls, the author has begun with "Morte d'Arthur," in effect preparing his implied national audience for the later reversal. The behavioral systems emphasized in "Coming" are mating, kin relations, and social relations. Arthur is successful both in surmounting the difficulties associated with his uncertain parentage and in winning the beautiful princess with whom he has fallen in love. In terms of social relations, he succeeds in building coalitions and achieving status.

The second, third, and fourth idylls have prototypically heroic and romantic, comedic structures as well. In "Gareth and Lynette," which I described earlier as a kind of "romantic fantasy," Gareth proves his worth by enduring his menial position in Arthur's kitchen before achieving knighthood and of course winning the hand of the demanding, sharp-tongued Lynette. Kinship is also a significant factor here: against the wishes of his mother, Gareth is identified as Arthur's nephew, and Arthur, who knows Gareth's identity but keeps it secret from others, is in essence supervising the initiation of his kinsman into knighthood. Another sort of courtship is narrated in "The Marriage of Geraint." Geraint is already an accomplished and handsome knight when he meets the meek and humble Enid, who lives in poverty. Of course, Geraint's heroism is focused on liberating Enid's family from the power of the man who has oppressed them and restoring their high social status, but his respect for Enid and her family in their humbled state and his insistence that she wear her old, unfashionable dress in accompanying him to Camelot suggest a transcendence of materialistic values related to social status that Tennyson probably associates with Christianity. "Geraint and Enid" is a continuation of their romance, but in a way that intertwines a heroic plot with the psychological drama of a marital relationship. Enid, ironically, is critical of her husband's neglect of his position as knight due to his exclusive devotion to her, and he misunderstands her motives. Nevertheless, their reconciliation is achieved by a kind of ritualistic violence in which Geraint performs knightly heroism in a formulaic way. After this resolution of marital difficulties, Geraint with his wife, though remaining in their own land, distant from Camelot, is once more focused on his responsibilities within the Arthurian coalitions.

The fifth idyll, "Balin and Balan," was discussed in some detail above, but here I want to reiterate the fact that this tragic story of brother killing brother is the turning point toward tragedy in the master narrative, and Tennyson not only focuses on the dysfunctional marriage union between Arthur and Guinevere and its effect in weakening coalitions vital to Arthur's Order but also on the basic ideas of survival and kin relations in particularly graphic ways. The horror of killing and death in this particular context is intensified. In "Merlin and Vivien," the sixth idyll, the victim is a key participant in the King's rise to power and his role as "magician" incorporates his function as storyteller. Merlin represents a center of cognitive activity in Arthur's kingdom and his destruction is of course a major blow. The implicit romance in the title and narrative structure of the seventh idyll, "Lancelot and Elaine," is made impossible by the illicit union of Lancelot and Guinevere, and the disruption of desirable mating patterns within the social order is made apparent.

In "The Holy Grail," the disintegration of Arthur's Order continues as his knights pursue imaginary visions of the Grail, and coalitions become further unraveled, but this eighth idyll with its emphasis on ideology and misguided spiritualism represents a kind of interlude in the series of tales featuring misguided and corrupted mating patterns. The emphasis on dysfunctional courtship and mating returns with the ninth idyll "Pelleas and Ettare," in which the perverted heroism of Pelleas is intimately related to the dishonorable mating patterns represented first by the knight Gawain, who uses his supposedly honorable mission in aid of Pelleas's courtship of Ettare as a pretext for mating with her himself and ultimately, of course, by the secretive union of Lancelot and Guinevere. As discussed earlier, the disillusioned Pelleas is transformed into the savage Red Knight, a kind of anti-Arthur, and in this narrative the intimate relationship between social relations and mating practices is clear. Then, in "The Last Tournament," the 10th idyll, the insidious love relationship between Lancelot and Guinevere is ironically paralleled by the love of Tristram for Isolt, where adultery is again an issue. The controlled violence of knighthood essential for conquest of the heathen and maintenance of Arthur's social Order has been perverted into internal, "self-destructive" violence that will literally lead to civil war. In the 11th idyll, "Guinevere," Arthur's sorrowful confrontation with Guinevere in the convent emphasizes the centrality of adultery, and in more general terms, the violation of rules and rituals associated with mating, resulting in the dissolution of this version of civil society, and the essential links between mating practices and social relations are made clear.

The great battle in the final idyll, "The Passing of Arthur," of course closes the tragic master narrative of Arthur and his kingdom. Arthur's individual life story represents the life story of a society, or even a civilization, and the tragic "death" of that civilization, obviously related to disrupted mating patterns, emphasizes not only their essential tie to social relations but issues involving kin relations and parenting. Ultimately, the issue is survival: the kingdom and the vast majority of ancillary characters in the narrative are dead, but in a coda, Arthur himself achieves a divine home in a version of the spiritualization found at the end of many prototypical narratives about heroes.[29]

Arthur temporarily triumphs over tribalism and helps to spread Christianity and the peaceful ideals of Christian civilization. Defeating a series of enemies, he builds coalitions that secure his capitol of Camelot. He promotes chivalry and related social institutions that serve to control male on male violence and offer protection to women. Nevertheless, he succumbs to fundamental problems related to mating, parenting, and kin relations, and his own masculinity is at the heart of these problems. An analysis of masculinity in the *Idylls* invites a further re-evaluation of Tennyson's great narrative poem and reveals structures of meaning in the work that have not yet been adequately discussed and imply a greater significance, a nightmarish prophetic quality that can be compared to that of Joseph Conrad's *Heart of Darkness*.

---

[29] See Hogan's table, "The Common Form of Prototypical Narrative Genres" (232–3).

Chapter 3
Barrett Browning's Construction of Masculinity in *Aurora Leigh*

One of the effects of the development of feminist scholarship within Victorian studies in the last quarter of the twentieth century was to restore critical interest in Elizabeth Barrett Browning's unique long poem *Aurora Leigh* (1857). Very popular in its own day, this Victorian classic suffered from the reaction to "Victorianism" at the end of the nineteenth century and then nearly disappeared from English literary history in the twentieth century before the rapid growth of gender studies that began in the 1970s made it inevitable that a groundbreaking story of a "woman poet" by the most famous female poet of Victorian England would receive attention. The revival of both the poet and the poem has not been without controversy, with some critics seeing her role as a traditionalist who defended "patriarchal" values,[1] but most celebrating Barrett Browning as a feminist who carved out a new cultural and aesthetic space for women.

Although some contemporaries praised *Aurora Leigh* extravagantly, John Ruskin, referring to it as the greatest poem in the English language,[2] reviews were mixed, and the Athenaeum perceptively pointed out that the poem would be "To some ... so much rank foolishness, – to others almost a spiritual revelation."[3] Modern feminist commentators typically analyze the contemporary critical response to show that the representations of women and women's experience in the poem were consistently at issue in both favorable and unfavorable evaluations. "Gender issues" were central to Barrett Browning and her original audience as well as to today's readers. For obvious reasons the portrayal of women's roles, beginning with Barrett Browning's concept of the feminine poet, has been the primary focus of gender-based studies of *Aurora Leigh*. Nevertheless, Barrett Browning's representations of masculinity in the poem are complex and distinctive,

---

[1] For example, Deirdre David in *Intellectual Women and Patriarchy* (1987) emphasizes Barrett Browning's supposed complicity in supporting male hegemony and bourgeois culture. Taking a middle ground, Sueann Schatz, in a 2000 article, explains how the philosophy of her heroine Aurora Leigh "is a complex weaving of newly-emerging feminist and long-embedded patriarchal ideas" (103).

[2] Ruskin wrote, "Mrs Browning's *Aurora Leigh* is, as far as I know, the greatest poem which the century has produced in any language" (*Things to be Studied*, 1856). Cited in Deirdre David, *Intellectual Women and Victorian Patriarchy: Harriet Martineau, Elizabeth Barrett Browning, George Eliot* (London; Macmillan, 1987), 95. See Taplin, 311.

[3] Cited in Helen Cooper, *Elizabeth Barrett Browning, Woman and Artist*, 149.

particularly in her portrayal of the character Romney Leigh. Because most modern commentators on the poem have assumed that one of the poet's primary goals was (or should have been) to liberate herself from confining and limiting "patriarchal" traditions, critical interest in her treatment of masculinity has been focused almost entirely on that issue.

This is a case where gender and genre issues are inextricably combined. As Helen Cooper puts it, "The male epic tradition and the female novel form her voice."[4] The poem is composed of nine books and has been described variously by Barrett Browning and her critics as an epic, a verse–novel or novel–poem, a *Bildungsroman* (and *Kunstlerroman*). As Dorothy Mermin pointed out, Barrett Browning, with aspirations well beyond those of "poetesses" such as Joanna Baillie, Felicia Hemans, and Letitia Landon, was both frustrated and liberated by her role as "the first woman poet in English literature,"[5] and I want to acknowledge the important work of Mermin, Cooper, Cora Kaplan, Joyce Zonana, Rebecca Stott, Marjorie Stone, and other scholars who have explored in considerable depth Barrett Browning's sense of her revolutionary and problematic status as a woman writer, her concern for women's issues, and her role as a female sage and social critic.

Barrett Browning, who shared tutors with her brother Edward ("Bro"), began to study Greek at the age of 11 and was well acquainted with classical epic traditions. She had a lifelong fascination with the genre of epic. Her first major poem was *The Battle of Marathon* (1819), a four-book epic. She maintained her interest in classical literature and published, for example, a translation of Aeschylus's *Prometheus Bound* in 1833. Although most modern critics assume that her interest in Greek epic traditions and classical literature in general belongs to an early period of apprenticeship, before she found her "authentic" voice, some, like Simon Avery and Rebecca Stott, noted that even the early *Battle* project implies an interest in the re-emergence of ancient Greek democratic ideals that were part of the nineteenth-century movement to establish a modern, independent Greek state, and in her 1833 version of Aeschylus, they find analogues to contemporary British political issues, including a reaction to the abuses addressed by the Reform Act of 1832.[6] Furthermore, as pointed out by Evelyn Hanley, Barrett Browning's continuing high regard and affection for classical tradition is verified by her *Essays on the Greek Christian Poets and the English Poets*, published posthumously in 1863.[7]

In any case, in 1844, when Barrett Browning, still living in London, began to think about the possibility of writing a long narrative poem, she had in mind not an imitative, backward-looking book but an experimental, contemporary one that would capture the spirit of the age in which she lived. Her short poem

---

[4] Cooper, 147. Marjorie Stone discusses the reception of *Aurora Leigh* by at least some mid-Victorians as "a modern epic" (142–5). Among the terms that Barrett Browning herself used to describe the poem is "poetic art-novel" (Letters of EBB, II:228).

[5] See Mermin, 1. On the question of genre, see Mermin, 184.

[6] See Avery and Scott in *Elizabeth Barrett Browning*, 45–7, 54–6, 61–4.

[7] See Evelyn Hanley, 37.

"The Cry of the Children," published in *Blackwood's Magazine* in August 1843, had demonstrated her willingness to confront openly controversial social issues of the day, in this case the deplorable conditions of young people working in factories. Writing "Lady Geraldine's Courtship," included in her 1844 collection, encouraged her to pursue other narratives of "modern love," and this poem, which contains an allusion to Robert Browning's poetry, became associated with her own real-life courtship by her future husband, who introduced himself in a letter to her after reading the 1844 *Poems*. Toward the end of 1844, she suggested in a letter to Mary Russell Mitford that Byron, whom she had long admired, was one of her models: "I want to write a poem of a new class, in a measure – a *Don Juan*, without the mockery and impurity … and having unity, as a work of art, – & admitting of as much philosophical digression (which is in fact a characteristic of the age) as I like to use" (*Brownings' Correspondence*, 9:304). In 1845 she began writing the remarkable series of new sonnets from the point of view of a *female* poet–lover that would make up the substantial sequence *Sonnets from the Portuguese*, presented to her husband Robert Browning in 1849 and included in her new edition of *Poems* (1850).

In what is surely the most prominent love story in British literary history, she and Browning had begun their correspondence in 1845, subsequently developing a romantic relationship and then getting married and moving to Italy in 1846. By the spring of 1847, they had moved to Florence and, in the following summer, to their celebrated lodgings at Casa Guidi, where Barrett Browning began to write the long poem *Casa Guidi Windows*, expressing her intense interest in contemporary Italian politics, her idealistic support of the *Risorgimento*, and a generally positive attitude toward Italian (especially as compared with English) culture. Eventually published in 1851, this long poem makes references to Michelangelo's Medici Tomb sculpture of Aurora, goddess of the Dawn[8] and anticipates the physical and cultural setting of *Aurora Leigh* and the double – Italian and English – identity of its heroine. In the meantime, Robert Wiedemann (Penini or Pen) Browning was born in March of 1849, and Barrett Browning devoted herself to the care of her son while continuing to write poetry. When published, her new book would be her first publication not dedicated to her father; instead, it is dedicated to her distant cousin, John Kenyon, a wealthy bachelor and patron of the arts, who happened to be a friend of the Browning family and had assisted Robert in preparing for his first visit to Elizabeth's house. Then he had bestowed an annual gift on Elizabeth after the birth of Pen and on his death in 1856 left a bequest that made Robert and Elizabeth financially independent.

She started working on *Aurora Leigh* in 1853, and the poem appeared in November 1856. Its nine books (associated with the nine books of the Cumaean Sybil and perhaps the nine months of a woman's pregnancy, in contrast to the traditional 12-book structure of the male epic used by both Tennyson and Browning) on the one hand present a tightly structured plot but on the other hand

---

[8] See Barrett Browning's *Casa Guidi Windows*, I, 73–4.

imply uncertainties and questions about the overall meaning of Aurora's narrative. The first five books take Aurora from her childhood in Italy to her adolescence and young adulthood in England and, finally, her decision to return to Italy. In the first book Aurora introduces herself as a young author and tells the romantic story of her English father's marriage to his Florentine wife, the mother's early death (when Aurora is four), and the father's attempt to assume the roles of both father and mother, the father's death (when Aurora is 13), and the orphaned Aurora's move to England to live with her father's sister. I return to the character of the father, who plays a crucial role in Aurora's life and is one of the two most important male characters in the poem whose masculinity is central to my discussion.

Aurora's implicit social criticism of English society, associated with her developing poetic sensibilities, begins with her unhappy life in the home of her aunt, as described toward the end of the first book, and continues in the second book, which features Aurora's commitment to her vocation as poet and rejection of her cousin Romney Leigh's marriage proposal on her 20th birthday. Selections containing the scene in which Aurora rejects Romney's marriage proposal – along with his ideas about gender and poetry and his general world view – are the ones most commonly anthologized and quoted in academic literary textbooks today, and of course I plan to discuss Barrett Browning's representation of Romney's masculinity and his role in Aurora's narrative. The aunt dies, leaving Aurora, age 20, with very limited resources. Romney attempts to give her money by pretending to have made a monetary gift to the aunt just before her death, one that Aurora would have been entitled to inherit. Aurora refuses to accept the money, however. In the third book, an independent Aurora, living in a London garret at the age of 27, is successfully developing her craft as a writer, and other major characters are introduced. Marian Erle is a young working-class woman who has suffered from an abused childhood. Romney helps her to escape from her dysfunctional family and then, in his misguided idealism, plans to "rescue" her definitively by marrying her. The elitist Lady Waldemar, who emerges as a villainess described in lamia or serpent imagery, asks Aurora to help prevent this union (which the reader, through Aurora's commentary, is encouraged to think of as wrong, although the Lady's motives in opposing it are entirely selfish, based on her own desire for Romney). In the fourth book, we learn more about Marian's life story, and then, in an embarrassingly public scene, Romney is left standing at the altar and the marriage with Marian does not take place. The fifth book is transitional, exploring Aurora's insights about the function of art in human life (this is where she argues against Tennyson's supposed medievalism, in contrast to her own modern epic), describing a party at the home of the aristocratic Lord Howe with its assortment of English and European intellectuals, and concluding with Aurora's decision to return to her Italian homeland.

The development of the *Bildungsroman/Kunstlerroman* narrative in this initial group of five books is, in some ways, straightforward, although, of course, here, in a gender reversal, the artist is a woman rather than the traditional male artist. More complex and difficult to analyze is the introduction of a kind of double for the artist, or second protagonist. From the beginning, readers of *Aurora Leigh* have identified the eponymous heroine with the author, and both positive and negative assessments of the poem have taken into account this obviously close and intentional association. However, the physical description of the low-born Marian Erle (unlike Aurora's) is much like that of the author – for example, Marian's curly black hair is described as spaniel-like, recalling Barrett Browning's own curls that were often compared with the fur of her beloved pet dog Flush, and Aurora's close identification with Marian, along with Romney's proposal to both women and other factors, strongly suggest a kind of double-identity here. Another complication is in the narrative structure of the first five books, which the reader understands to have been written by Aurora in England prior to her journey back to Europe. This is more than a plot detail: the point of view of the narrator is not that which will be reached by the end of the story, dramatically transformed from the earlier one. That is, at the end of the first five books, the narrator – with no hint of contradiction from the implied author – apparently assumes that Aurora and Romney have parted forever. There is no foreshadowing of the reconciliation of the lovers at the end.

The rest of the poem is narrated in a kind of journal format, and books six and seven are written "on the road" in Paris and then in Florence. In the sixth book, Aurora meets Marian, accompanied by her illegitimate infant son in Paris, hearing, first, Marian's account of the abortive wedding – with her explanation of why she could not allow Romney to rescue her – and then an account of her journey to France, where she was accompanied by a woman acting as an agent of Lady Waldemar, who deceived her into thinking they were traveling to Australia and then delivered her to a French brothel, where she was raped. At first, Aurora's response to Marian's rape and resulting pregnancy and motherhood of her illegitimate son is condemnatory in a stereotypical way, but she gradually shifts to a more sympathetic view of Marian as victim and an appreciation of her loving devotion to her son in spite of his origins. After Marian concludes her narrative in the seventh book, Aurora offers to take her and her child to Florence, and the three arrive there together. The reader is informed that letters written by Aurora and Romney do not reach their destination, so communication between the two is severed. Nevertheless, in the eighth book, Romney arrives at Aurora's Italian home during an evening when Marian and her son are away. Incredibly, Aurora does not realize that Romney is now blind, as the two engage in an all-night conversation about art and society. In the concluding ninth book, Aurora reads the letter he has brought from Lady Waldemar, a resentful and envious letter that nevertheless acknowledges that, contrary to Aurora's assumptions, Romney has not married the Lady. Marian joins Aurora and Romney, and he once again proposes to her, but once again she refuses, thanking and praising him for his offer but explaining that her traumatic experience has made it impossible for her to love a man and that she will devote herself entirely to her role as mother. Marian retires and Romney's blindness is revealed. He tells Aurora the story of how he had converted Leigh Hall, his family home, into a Fourier-style phalanstery and how it was burned and looted by the people he had tried to help, including Marian's father, who caused the injury that resulted in Romney's blindness. Understanding that they are now free to unite, Romney and Aurora acknowledge their mutual love, and each confesses previous personal shortcomings and misunderstandings as they look forward to their forthcoming union, which is framed in Apocalyptic imagery taken from Revelation 21. In this ending, life and love are emphatically elevated above art and fame, and this is troubling for modern critics who want to interpret the affirmation of Aurora's personal independence and authorial success as the most fundamental values implicit in the text. For example, Alison Case refers to a "double teleology for the novel [sic], in which the struggle toward artistic independence and success could be kept from the undermining influence of the traditional love story" (32). The problem is that it is the love story at the end that the reader is asked to accept as the overarching, unifying narrative. This is not to say, however, that Aurora's story is thus reduced to a simplified, stereotypical tale of love and marriage; on the contrary, the ending and the overall structure of the poem are complex and full of uncertainties in spite of the celestial images associated with God's truth, and I examine some of these uncertainties in the course of my discussion.

For obvious reasons, the character of Romney Leigh is of major interest in any analysis of masculinities in *Aurora Leigh*, but Aurora's father is also an important male character, one who cannot be altogether ignored by critics because of his close relationship with his daughter as portrayed in the first book (and because of the temptation to identify him with Barrett Browning's own father, who disinherited her after she eloped with Robert Browning)[9] – but who deserves much more attention than he usually receives. The father's legacy is fundamental to Aurora's sense of belonging to a family unit in her childhood, an affectionate bond that opens her up to a love of the world in a larger, spiritual sense. It also includes a love of learning, a scholarly background in the classical humanist tradition, and a critical sensibility that prepares the way for Aurora's critique of British society.

This last component of Aurora's heritage from her father is often ignored by critics, but it is extremely important. By traveling to Italy and falling in love with an Italian woman, the father distanced himself from his ancestral English roots. When the orphaned Aurora travels to the Leigh estate to take up residence, her maiden aunt, who assumes the parental role, resents the Italian, maternal component of her identity and by implication the brother who "deserted" his family background and married a foreigner. The narrator Aurora begins the first book with an account of her parents' courtship and marriage. In a rather melodramatic incident in Florence, where he had come to spend a month "studying DaVinci's drains," a young man

---

[9] When Edward Moulton-Barrett initially became suspicious of Browning's visits, he forbade Elizabeth from leaving the house. After her elopement, he disinherited her and spoke of her as if she were dead. He did not open the letters she wrote to him afterwards, and there was no reconciliation. The story of Barrett Browning's estrangement from her father is, of course, well known and was incorporated into the popular legend as represented in Rudolph Bessier's 1931 play *The Barretts of Wimpole Street* and elsewhere.

falls in love (at first sight) with a young woman he observes in a religious street procession. This "austere Englishman / ... after a dry life-time spent at home / In college-learning, law, and parish talk, / Was flooded with a passion unaware" (I, 65–8). It is true that Aurora draws a contrast between her mother and her father, who was not magically transformed into a gregarious, emotional Italian: he remains an "austere" Englishman, but one who has emigrated to a new home and found a place in a different culture.

The mother dies when Aurora is four, and the father, as a single parent, in spite of his devotion to his daughter, cannot take the place of the dead mother. The assumption here is that women with children have an innate capacity to deal with the young:

> Women know
> The way to rear up children, (to be just)
> They know a simple, merry, tender knack
> Of tying sashes, fitting baby-shoes,
> And stringing pretty words that makes no sense,
> And kissing full sense into empty words. (I, 47–52)

Fathers "love as well" but "with heavier brains" (I, 61). Sensing his inadequacy to take the place of Aurora's mother after her death, he decided to leave the city of Florence and move with his daughter to "the mountains above Pelargo, / Because unmothered babes, he thought, had need / Of mother nature more than others use" (I, 111–13). In this natural setting, the young Aurora lives with her father for 9 years; she is 13 when he dies: "His last word was 'Love,–' / 'Love, my child, love, love! – (then he had done with grief) / 'Love my child.'" (211–13). The final words of the dying father have a far-reaching significance in the poem. On the most obvious level, they express the father's undying love for his departed wife and for his daughter, the central force and source of meaning in his personal life, and at the same time they express a gentle command in the biblical tradition of "Beloved, let us love one another: for love is of God" (1 John 4:7–10), expanded by the philosophy of Emmanuel Swedenborg, who influenced Barrett Browning's concept of God as infinite love and infinite wisdom. Through "correspondences" between the seen and the unseen, God's spiritual truths are embodied in the physical world, and the most important correspondence here is that between "the delights of heaven" and "the delights of conjugal love."[10] In the context of the father's final words, the meaning of love is broad indeed, and these words prophesy both Aurora's vocation as a seer poet whose art is informed by a love of God, mankind, and the natural world and her loving union with Romney Leigh, in the final book of the poem, when she acknowledges that "Art is much but love is more ... /Art symbolizes heaven, but Love is God / And makes heaven" (IX, 656–9). And it is significant that, after all the impassioned expressions of love for each other repeated by Aurora and Romney near the end of the poem as they face their future

---

[10] See Avery and Scott, 136–8.

together, it is Romney who declares that Aurora will "work for two" while he "for two, shall love!" (IX, 911–2). This apparent reversal of traditional gender roles at the end of the poem is prepared for by the father's association with love in the first book.

Memories of her lost beloved father sustain Aurora as she struggles to adjust to her new life among his relations in England. His legacy consists of both love and wisdom, which is associated with the books he left her: as she reviews the Latin and Greek classics she had studied with him, "What my father taught before / From many a volume, Love re-emphasized on the self-same pages." (I, 711–2). Then she discovers in her new home a "garret-room" piled with cases of his books, so that his posthumous influence on her education continues. She reads one book after another almost indiscriminately, but "when the time was ripe," she begins to focus on the poets, God's "truth-tellers," who would inspire her to develop her own sense of poetic vocation.

Aurora clings to the memory of her father as she reads his books and becomes both an intellectual and artist in her own right. The father does not merely pass down a traditional, classical body of learning to be accepted without question. His critical attitude is also an essential part of Aurora's inheritance:

> He taught me all the ignorance of men,
> And how God laughs in heaven when any man
> Says 'Here I'm learned; this, I understand;
> In that, I am never caught at fault or doubt.'
> He sent the schools to school, demonstrating
> A fool will pass for such through one mistake,
> While a philosopher will pass for such,
> Through sad mistakes being ventured in the gross
> And heaped up to a system.
>   I am like,
> They tell me, my dear father. (I, 190–99)

The critical bent in Aurora emerges early, as she rejects the stereotypical learning and conventional wisdom associated with her aunt, and of course it is as criticism of contemporary British Victorian society that Barrett Browning's portrayal of the mature poet Aurora Leigh is most highly valued by literary critics today. But this nonconformist, independently minded personality self-consciously identifies herself as being in the tradition of her father. More subtly, the father's rejection of overly systematic thinking prepares the way for Aurora's rejection of Romney's socialist ideology. It is only when Romney sees past his theoretical, systematic thinking to acknowledge the primacy of *love* that Aurora is ready to accept him as her mate. The reversal of the heart (female) and head (male) stereotypes at the end of the poem is anticipated by the central "message" of Aurora's father. For Barrett Browning, there is in the heritage of Christian humanism, however patriarchal, a powerful critique of prideful, systematic thinking. The poet–prophet, attuned to God's truth, is superior to the philosopher with his abstractions.

There is a marked bias here in favor of Barrett Browning's Evangelical Christian heritage and against Enlightenment thought, and when she asserts her own kind of feminism in developing her concept of the female seer poet, it is a feminism that draws on woman's morality and her special proximity to God's truth rather than the systematic, rational ideal of gender equality developed by John Stuart Mill in the tradition of Mary Wollstonecraft.[11] One expression of her ideal of the visionary and prophetic poet is found in her poem "A Vision of Poets" (1844).[12]

In her English isolation, the young Aurora "read much. What my father taught before / From many a volume, Love re-emphasized / Upon the self-same pages: Theophrast / Grew tender with the memory of his eyes,' / And Aelian made mine wet" (I, 710–14). Later, she must sell her father's books to finance her return to Italy, though she cannot bear to part with one by the Greek philosopher Proclus because she associates it with the memory of the time her father scolded her for pressing an iris flower inside it, staining the pages: "Ah, blame of love, that's sweeter than all praise / Of those who love not!" (V, 1241–2). The memories of her father and mother become increasingly dreamlike and remote, and when she returns to Italy she is unsuccessful in her attempt to re-establish an emotional link to her childhood home, but the heritage of love and learning – indissolubly linked together – is a permanent gift from her father and provides the foundation for her intellectual and aesthetic vocation.

As suggested earlier, Aurora's rejection of Romney's marriage proposal in the second book of *Aurora Leigh* is today the best known and most celebrated scene from the poem. When it appears in anthologies of Victorian literature, sometimes there is a footnote reference to the reformed Romney of the eighth and concluding ninth book. What I want to point out here, however, is that the unreformed Romney of books 2–7 is already an extraordinarily heroic figure, however misguided and confused in his idealism. His proposal to marry his poor cousin is framed by his devotion to his work, and Aurora's refusal focuses on this: "'Why sir, you are married long ago. / You have a wife already whom you love, / Your social theory'" (II, 408–10). In the conclusion, the mutual misunderstanding here is revealed to the reader as both characters are finally enlightened and apologize (profusely) to each other. Romney did not understand that Aurora's own work – poetry – is finally more effective in reforming society than his own theories and political activism;

---

[11] As Hanley puts it, "one of the common bonds she had with Browning was the belief in soul, the conviction of immortality, which was as deep-entrenched in her sense of the universe as in his" (50).

[12] Stephanie L. Johnson argues that "A Vision of Poets" is subversive of the male tradition of poetry. In her reading, "The Romantic seer with his visionary power, with his sacrificial vow that can bring apocalyptic upheaval to the heavens and earth, is framed in this poem – put in his place, where he dies. The female controls the visionary and apocalyptic forces for her own ends" (440). Nevertheless, according to Johnson, Barrett Browning leaves behind her earlier radicalism when she composes the apocalyptic vision at the end of *Aurora Leigh*. From Johnson's point of view, the female poet there gives way to male, patriarchal tradition.

and, on her side, Aurora did not realize that Romney's desire to have her join him in his life's work – to which he was totally committed – grew out of his sincere love for her: what he envisioned was a loving partner rather than a subservient assistant.

After being rejected by the woman he loves, Romney redoubles his efforts at reform. Seen from his own "enlightened" point of view as expressed in his narrative to Aurora in the eighth and ninth books, his efforts were misguided, based on the fundamental misunderstanding that the needs of the poor are entirely material – that providing them with the basic requirements of food and lodging will enable them to change their lives. In this Romney had been too much of his age: "'We're too materialistic, – eating clay'" (VIII, 630), "'And fumbling vainly therefore at the lock / Of the spiritual'" (VIII, 659–60). At the end of the poem, he is able to contrast his materialism with Aurora's poetic insight that people require meaning and purpose in their lives, to look beyond the natural world to the spiritual. To work for God exclusively in terms of a supposed good in the natural world according to manmade systems is sure to fail:

> 'Fewer programmes, we who have no prescience.
> Fewer systems; we who are held and do not hold.
> Less mapping out of masses, to be saved,
> By nations or by sexes. Fourier's void,
> And Comte is dwarfed, – and Cabet, puerile.
> Subsist no law of life outside of life;
> No perfect manners, without Christian souls:
> The Christ Himself had been no Lawgiver
> Unless He had given the life, too, with the law.' (IX, 865–73)

Romney has been converted to his new, spiritualized, worldview, by Aurora's poetry. Not just his own program and system but all varieties of socialism and materialist philosophy are doomed to fail. Aurora's alternative "spiritualism" aimed at human souls rather than bodies is associated not only with a loosely formulated Christian theology and the Swedenborgian system of correspondences previously mentioned, but also the tradition of the Romantic seer poet who finds a transcendent God beyond the natural world and Victorian sage literature like that of Thomas Carlyle, which also describes a sense of spiritual fulfillment that can be sought by meaningful work as a solution to the Condition of England Question.[13] In each case, the individual must follow the lead of his or her own soul, not an intellectualized, materialistic formula or philosophical system, however well intentioned. The "poetic" vision of the world to which Romney has been converted (by Aurora) is by implication a fully androgynous one, unlike the implicitly masculine systems of thought that rely on abstract formulations and statistics. Also excluded from this vision is a serious consideration of natural science and the prominent intellectual and religious issues of the day associated

---

[13] On evidence of Carlyle's influence in *Aurora Leigh*, see Avery and Stott, 192–7.

with evolutionary theory. (Nature imagery – especially regarding the Italian countryside – is generally described in feminine terms, suggestive of a "mother nature" created by God, and Barrett Browning's focus on the female body and spontaneous emotions can be compared with the techniques used by male Spasmodic poets, as discussed below.) Of course, the character Romney, in spite of his extensive confessions of error and expressions of self-humiliation in the concluding books, has been represented by Barrett Browning as a good man with the potential for spiritual growth and enlightenment. The reader can hardly take Romney's extensive professions of personal failure at face value; such a failure would not be an acceptable characterization of the heroine Aurora's lover. In fact his extreme humility is an admirable trait that is consistent with his history of self-sacrifice for others (individuals like Marian as well as the poor as a class and society as a whole).

As pointed out earlier, Romney is from the beginning of the narrative a Christian who has personally invested in the model of Christ; nevertheless, his failure has been in the intellectual misunderstanding that led him to embrace a materialistic philosophy of life. Instinctively generous and committed to helping others, he provides genuine assistance to Marian at the beginning of their relationship, and his reform efforts are motivated by the best of intentions. Furthermore, the early systematic, egotistic Romney has many positive masculine traits consistent with the normative values of Victorian bourgeois respectability, as discussed in previous chapters: he is stoic, dedicated, and hard-working, with a strong sense of duty. Beyond this, he has a strong sense of social justice, though his stubborn impracticality conforms to certain stereotypes of the male intellectual. Romney's conversion to Aurora's unsystematic, spontaneous moralism, which emphasizes the value of enlightened attitudes and contact with individual human beings rather than types or classes of people, poetry rather than planned work projects, is necessary before he becomes an appropriate mate and work partner for Aurora, in line with the values passed on by Aurora's father.

He is placed in social contexts where the reader is invited to compare him with other male characters, most notably at the evening party at Lord Howe's described in the fifth book, where Aurora uncomfortably joins the "talkers." Lord Howe himself is an MP who "expounds his theories / Of social justice and equality" (V, 593–4) while his wife listens with admiration. Unlike Romney, he is comfortable with his elitist social position and enjoys the political and philosophical discussions as a kind of sophisticated verbal game. In a conversation with Aurora, he attempts to pass along to her a letter from John Eglinton, whom Aurora identifies as an aristocratic landlord from an old family who has had previous romantic attachments with female celebrities such as actresses and artists, and who now apparently is interested in cultivating a friendship with a popular poet. Aurora refuses to accept the letter and also rejects Lord Howe's adage that "A happy life means prudent compromise" (V, 922). Although Howe, with his compromised ideals, is a superficially pleasant and agreeable authority figure, the hypocrisy of his position is emphasized by its juxtaposition at the party with the more egregious example of the discredited and spiteful Lady Waldemar. She was introduced in the third book as Marian's rival, an elitist who thought it wrong for Romney to align himself with the working class. Now she talks of her charity work, assisting Romney at his Leigh Hall phalanstery, and Aurora learns at the party that she is widely assumed to be Romney's "disciple" as well as future wife. Other male figures at the party include the smug Catholic Sir Blaise Delorme, who in his "gentle arrogance" feels morally superior to the others, and a German student, who is bristling with ardent (atheistic) political activism but assumes that Leigh is "our ablest man," despite his absurd clinging to anachronistic Christian belief along with this socialism. Sir Blaise, secure in his belief in the "catholic, apostolic, mother-church" cannot accept "Christian-pagans."

Aside from the plot complications involving Romney's supposed relationship with Lady Waldemar (to which I return below), what emerges in these scenes at Lord Howe's party is a contextualization of Romney's distinctive moral character and social reputation. As a member of parliament with an activist social and political agenda, he is a leading figure whose passionate commitment to social change is inherent in his personal life, but he retains a Christian belief in God. He accepts Fourrier's ideal of equality between the sexes along with radical plans for the redistribution of property but, consistent with his religious beliefs, rejects the concept of free love. Thus, even at his most radical stage, before turning his life around in response to Aurora's poetic vision, he is a man of integrity, focused on deeply held personal beliefs that transcend his own life and uninterested in compromises that might win him social approval. If he later disavows his "materialistic" philosophy, his "materialism" has not been directed toward his own personal gain or comfort, and his fundamental change of position is the result of his own interpretation of Aurora's poetry, not a response to social pressure.

It is worthwhile to focus on his relationship with Marian Erle in more detail to show how this social reformer in fact plays what can be described as a knightly, chivalric role in her rescue. Although his "romance" with Marian Erle is based primarily on ideology rather than "true love," Romney does in fact rescue her from a degraded family environment in which she has been abused by a mother who attempts to sell her sexual favors: "there, a man stood, with beast's eyes … / And burning stertorous breath that hurt her cheek. / The mother held her tight, / Saying hard between her teeth – 'Why, wench, why wench, / The squire speaks to you now./He means to set you up, and comfort us'" (III, 1049–56).

Marian has suffered gravely as a victim of social and parental abuse, and though Romney's grand scheme of a marriage that will help to reform society by serving as a symbolic bond between a privileged upper class and the suffering, working class poor is based on foolish ideology (however heartfelt his commitment), it is in a sense a noble gesture that requires his personal sacrifice. Even after suffering the violent destruction of his property by rioters ungrateful for his reformist efforts and physical blindness as a result of an injury apparently caused by Marian's father, he maintains the commitment implicit in his original marriage proposal to Marian, who now is raising an illegitimate son – conceived as a result of her rape – and he renews his offer when he finds her with Aurora in Florence. Disillusionment in his socialist idealism does not affect his sense of honor and Christian morality. Responding to Aurora's incredulous expression "Ah – not married" after she has read Lady Waldemar's letter and discovered that Romney's supposed union with her insidious rival has not taken place, he declares: "'You mistake.' / 'I'm married. Is not Marian Erle my wife? / As God sees things, I have a wife and child; / And I, as I'm a man who honours God, / Am here to claim them as my child and wife'" (IX, 177–81). Both Romney's renewed proposal and Marian's rejection are heroic. In spite of her overpowering gratitude to Romney, Marian realizes that she is incapable of loving him (or any man) in a romantic sense: she is totally committed to her role as mother. Of course, she also understands that it would be wrong to take advantage of Romney's saint-like goodness – even to secure a father for her beloved son – not only because it would be impossible for her to feel a (conjugal) love for him but because it would be wrong to block the love relationship between Romney and Aurora that is "meant to be." When Marian refuses Romney, she addresses him as "master, angel, friend" and declares, "I know you'll not be angry like a man / (For you are none)" (IX, 353–4), affirming Romney's extraordinary masculinity.

Far from representing his willingness to deny his romantic feelings for Aurora, Romney's renewed proposal to Marian confirms his integrity and his identity as an exceptional male mating partner whose loyalty, when formally expressed, will be unquestioned. Romney functions as a remarkably heroic character, and, apart from the sensationally odd and melodramatic conclusion in which a blind Romney shares Aurora's spiritual vision, Barrett Browning's representation of this idealized mate for her heroine is complex. Romney's initial role in his relationship with Marian as "runaway" is that of a social worker. Sensitive to her tragic story of abuse, as a Christian socialist, he treats her with respect and charity. In the third book, Aurora reconstructs Marian's account: "She told me how he had raised and rescued her / With reverent pity, as, in touching grief, / He touched the wounds of Christ, – and made her feel / More self-respecting" (III, 1223–6). Then, "to snatch her soul from atheism, / And keep it stainless from her mother's face, / He sent her to a famous sempstress-house / Far off in London, there to work and hope" (III, 1228–31). Barrett Browning establishes this good work on the part of Romney, described primarily in terms of Christian charity rather than socialist ideology, prior to his well-intended but misguided plan to marry her in a spectacular public ceremony meant to counteract class warfare. Although he is sincere in his reformist ideas and in his affection for Marian, it is doubtful that he has a truly romantic attachment to her. His idealism toward this imperiled woman is not unlike that associated with Tennyson's Arthur, and I return to the idea of chivalry below.

In this way Romney is established as a sympathetic, caring person and a man of his word, the kind of man who is likely to pursue a long-term investment in a romantic relationship – when it comes along. At the same time, he exhibits some characteristics of an adventurous, even "Byronic" hero in his flouting of social conventions to pursue an idealistic cause. Unlike the superficial, frivolous, or flirtatious men at Lord Howe's party, he is motivated by profound intellectual and moral commitments, but he is not a boring prig. He does not share the aesthetic sensibilities of Aurora's sympathetic artist friend Vincent Carrington (who marries Aurora's enthusiastic follower Kate Ward), but he maintains a persistent masculine "otherness" that is potentially attractive. While retaining an unquestioned sexual morality (there is nothing of the "cad" in Romney),[14] his sexual attractiveness to women is certified by the passion of Lady Waldemar. As an heir to the Leigh family fortune, he is willing to share it (however unwisely) with the poor, but he retains his status as an English gentleman. The tragedy of his blindness and his blatant failure as a social reformer, since he remains morally pure, add to the pathos of his "Romantic" heroism and ironically make him acceptable as the husband of Barrett Browning's poet–heroine. Finally, the melodramatic circumstances of their union at the end of the poem overshadow the fact that for all her individualism and feminism Aurora has found a normative, heterosexual love relationship with a man she can respect and trust. Furthermore, by marrying her cousin she reaffirms her familial (and national) ties while maintaining her independence as an artist living abroad with reformist ideas and an emotional attachment to Italy. In the conclusion, following in the path prophesied by her dying father, Aurora acknowledges the primacy of love over poetry. At the same time, however, as already pointed out, it is the blind Romney who says, "Shine out for two, Aurora, and fulfil / My falling-short that must be! work for two, /As I, though thus restrained, for two, shall love!" (IX, 910–12).

We can think of Romney as a fulfillment of the Christian humanistic vision that Aurora associates with the memory of her absent father. But it is as if Barrett Browning cannot imagine the idealized relationship with Aurora unless Romney is literally, physically dependent on Aurora for his sight. Although generations of readers have noticed certain similarities between the "love story" plot of *Aurora Leigh* and that of Charlotte Brontë's *Jane Eyre*, Barrett Browning was reluctant to admit an influence, and she pointed out that Romney, unlike the character Mr. Rochester, is not disfigured or disabled beyond his blindness at the end.[15] Of course, the reader takes Romney's blindness to be a significant physical disability, and let us also recall the extraordinary significance of physical sight in the Romantic epistemology embraced by Barrett Browning and celebrated in *Aurora Leigh*. Representative passages are abundant, but this one is from John Ruskin's extended study of art and the natural world, *Modern Painters:* "The greatest thing a human soul ever does in this world is to *see* something, and tell what he *saw* in a plain way. To see clearly is poetry, prophecy, and religion, – all in one" (*Works*, V:333). In her essay "Blinding the Hero," Mary Wilson Carpenter reviews interpretations of Romney's "metaphorical" blindness and stresses what she sees as Barrett

---

[14] On the topic of "cads" or "men who consistently demonstrate a marked desire for sexual variety," see Daniel J. Kruger, Maryanne Fisher, and Ian Jobling, "Proper Hero Dads and Dark Hero Cads."

[15] See Mermin, 185 and 271–2, n. 4.

Browning's antipathy to conventional Victorian masculinity and the "manly" men portrayed in the novels of Charles Kingsley. This is not necessarily an indication of Barrett Browning's feminism because, as pointed out earlier, Tennyson was also uncomfortable with representations of "manliness" like those found in Kingsley's fiction, and I return to this comparison in Chapter 6, although the blinding of Romney is a melodramatic and in some ways problematic aspect of the poem. Carpenter suggests that Romney's blindness "seems to mark his *elevation* to the position of Aurora's equal, not his lowering" (57).

A more cynical assessment of the ending might be that only Romney's blindness ensures that his primary ambition in their future life together will be to support Aurora in *her* work (poetry) and that her identity will not be fixed by his "male gaze," though Barrett Browning surely had in mind positive models ranging from her blind mentor and teacher Hugh Stuart Boyd[16] to John Milton and the mythical Greek prophet Tiresias, and may have thought of the insights gained by the blinded Oedipus. At any rate, in spite of the elaborate celestial imagery at the end, the union between Aurora and Romney is described as sexual and erotic: "There were words / That broke in utterance ... melted, in the fire; / Embrace, that was convulsion ... then a kiss ... / As long and silent as the ecstatic night, – /And deep, deep, shuddering breaths, which meant beyond / Whatever could be told by word and kiss" (IX, 719–24).

This romantic climax is prefigured at the beginning of book eight, when upon his entrance Romney is associated with mythological images of a sea-king. Aurora, who has been reading Boccaccio in a dreamy mood, is looking at the river and fantasizing, just as night is falling:

> Some gaslights tremble along squares and streets;
> The Pitti's palace-front is drawn in fire;
> And, past the quays, Maria Novella's Place,
> In which the mystic obelisks stand up
> Triangular, pyramidal, each based
> On a single trine of brazen tortoises,
> To guard that fair church, Buonarroti's Bride,
> That stares out from their large blind dial-eyes,
> Her quadrant and armillary dials, black
> With rhythms of many suns and moons, in vain
> Enquiry for so rich a soul as his, --
> Methinks I have plunged, I see it all so clear . . .
> And, O my heart,.. . .the sea-king!

---

[16] The blind scholar and poet Hugh Stuart Boyd was an important influence on Barrett Browning as her friend and teacher, as confirmed by her diary entries during the period 1831–32. She served as his reader and amanuensis, and he encouraged her study of classical, especially Greek, authors. Later, she visited his house following her secret marriage to Browning in 1846, and she wrote commemorative sonnets to him after his death in 1848. On Barrett Browning's deep emotional attachment to Boyd, see Barbara Dennis, *Elizabeth Barrett Browning: The Hope End Years*, 82–4.

In my ears
The sound of waters. There he stood, my king!
(VIII, 48–62)

Just before this introduction of the image of the sea-king, "The duomo-bell / Strikes ten, as if struck ten fathoms down, / So deep" (VIII, 44–6), and this suggests the line "Full fathom five thy father lies" from Shakespeare's *The Tempest* (I.ii.397). Aurora, a mature artist who has returned to her childhood home, is making unconscious or half-conscious associations with her father – and perhaps Barrett Browning may have in mind her own beloved brother (who died by drowning) – as the transformed and romanticized Romney, now a potential lover, is being reintroduced.

It is useful to place Romney's extraordinary masculinity in the larger context of gender issues implicit in the poem. As suggested earlier, Lady Waldemar plays a key role in the "love and marriage" plot, and although her representation as an evil "lamia" figure, a serpent woman, may seem somewhat stereotypical, her function in the plot is more complex then that image suggests. To begin with the most obvious aspects of her role, she comes to be perceived by Aurora as her rival for the affections of Romney, but Aurora's critical opinion of her is established in the third book during their conversation when the Lady, in her brazen, unexpected visit, boldly declares her love for him and attempts to enlist Aurora's help in preventing what she considers to be the absurdly unsuitable union of Romney and Marian. Lady Waldemar's designs on Romney in fact lead to an important relationship with Marian as well. She advises Marian as a kind of surrogate mother figure, initially convincing her that she would not be an appropriate wife for Romney and then arranging for what is supposed to be Marian's emigration to Australia in the company of the Lady's former maid, a sinister woman who instead guides Marian to the French brothel where she is drugged and raped ("Twas only what my mother would have done" [VII, 9]).

Ironically, a wiser Marian – removed from the overpowering influence of Romney – eventually comes to understand that it would have been wrong for her to marry the man that she idolizes but does not love in a romantic way, but the manipulative Lady Waldemar's motives in this scheme, even if (as she claims in her final letter to Aurora) her agent acted independently in changing Marian's destination, are entirely selfish and not at all concerned with Marian's welfare. She simply wants to replace the younger woman as Romney's lover. As Aurora comes to believe that she has been successful, the reader sees the Lady increasingly through the heroine's eyes as a particularly negative representation of female sexuality, and of course she uses her aristocratic social identity as well as her physical attractiveness in a shameless manner. Aurora describes her strikingly attractive but indecent appearance at Lord Howe's, referring to her "alabaster shoulders and bare breasts": "If the heart within / Were half as white! – but, if it were, perhaps / The breast were closer covered and the sight / Less aspectable, by half, too" (V, 618, 623–6).

I am not implying that Barrett Browning is here displaying conventionally modest "Victorian" sensibilities about women's bodies. On the contrary, modern critics of *Aurora Leigh* often praise the poet's very positive and striking references to women's sexuality, including this often-quoted description of her mission as a modern epic poet:

> Never flinch,
> But still, unscrupulously epic, catch
> Upon the burning lava of a song
> The full-veined, heaving, double-breasted Age:
> That, when the next shall come, the men of that
> May touch the impress with reverent hand, and say
> 'Behold, – behold the paps we all have sucked!
> That bosom seems to beat still, or at least
> It sets ours beating. This is living art,
> Which thus presents, and thus records true life.' (V, 214–22)

Earlier, in defining her poetic work to describe the "multitudinous life" of God's world she had referred to the "strain / Of sexual passion, which devours the flesh / In a sacrament of souls" and "mother's breasts / Which, round the new-made creatures hanging there, / Throb luminous and harmonious like pure spheres" (V, 14–18).

Lady Waldemar's sexual immodesty and shameless lust for Romney are combined with her role as one of the poem's "bad mother" figures in manipulating the vulnerable Marian. In this she can be compared with Marian's real mother and contrasted with Marian herself. Significantly, Romney, despite the assumptions made by many, including Aurora herself, does not succumb to the seductive physical charms of Lady Waldemar. At Lord Howe's party, Aurora meditates on the difference between men and women in sexual matters: Romney, like other men, looks for *a* woman, *a* wife. Where "the man discerns /A sex ... / we see but one, ideally / And really: where we yearn to lose ourselves / And melt like white pearls in another's wine, / He seeks to double himself by what he loves, / And make his drink more costly by our pearls" (V,1075–81). Aurora's "cultural" stereotypes about gendered differences between men and women in their approaches to love and marriage are in fact related to biological differences discussed in the first two chapters.

The idea of "doubling himself" here of course goes beyond sexuality and supports the idea of Romney's seeing first Aurora and then Marian as merely a helpmeet to assist him in his own visionary work. But, as Aurora discovers in the end, Romney is different, capable of self-analysis and spiritual growth, and this is consistent with Marian's implication – in the long speech in which she rejects his renewed proposal in Florence – that he is somehow angelic, even godlike, not an ordinary man: "'O Romney! O my angel ... / *Thee* I do not thank at all: / I but thank God who made thee what thou art, / So wholly godlike'" (IX, 281–6). As already pointed out, however, this does not mean that his goodness, spirituality, and loyalty come at the price of his sexuality. Lady Waldemar finds him attractive even after his blindness – until he asks her to read to him from Aurora's new poem and his expressions of love for the poet finally drive her away and inspire her to write the bitter letter he delivers to Aurora. And the passionate kisses of Romney and Aurora at the end are described in physical terms that remind us of the Swedenborgian "correspondences" between spiritual and conjugal love.

The complex masculinity of Romney is especially apparent in his role as male "hero." Barrett Browning in some obvious ways undercuts and satirizes masculine chivalry in her poem.[17] When Aurora is disgruntled because she believes that Romney probably has married Lady Waldemar, she chides herself for not accepting Romney earlier and saving him from "Lamia": if God had made her like some women, "to save men by love, – / By just my love I might have saved this man" (VII, 185–6). Then she begins to cry but reacts against playing the part of the woman who weeps frivolously: "It seems as if I had a man in me, / Despising such a woman" (VII, 213–15). Deciding to take action to save her cousin, she declares, "The world's male chivalry has perished out, / But women are knights-errant to the last; / And, if Cervantes had been Shakespeare too, / He had made his Don a Donna" (VII, 224–7); and, although she knows she may be too late, she writes one letter to Lord Howe, informing him of Marian's fate as the result of Lady Waldemar's plot, and another to the Lady herself, charging her with the betrayal of Marian and sarcastically warning her – if already married to Romney – to be kind to him and not reveal her authentic, ignoble identity. The moral superiority of the stereotypical Victorian woman–angel is tied to her passivity, her potential victimhood, her need to be protected or saved by her knight, and Barrett Browning is rejecting this idea.

Earlier, of course, Romney had wanted to rescue Marian in a dramatic gesture that was to him based on socialist ideology and Christian morality but was also simultaneously chivalric, knightlike. Marian, in effect, would not allow him the opportunity to play the role of Perseus/St. George, and his physical blindness, along with Aurora's sense of self-achievement and physical security (contrasted with her emotional needs) at the end, precludes his playing this role in a traditional sense with Aurora, either. Instead, Aurora will care for *him*, lead *him*.

And yet, as discussed, Romney's relationships with both Marian and Aurora are not only generous but heroic, and his final loving union with Aurora is both physical and spiritual. To achieve these effects, Barrett Browning uses very conventional plot devices in unusual ways. The "love triangle" in this narrative, as previously suggested, involves a curious doubling in the role of heroine. The intense sense of "sisterhood" between Aurora and Marian (after Aurora has come to accept the other's "single motherhood" as not only blameless but admirable) unites the two: "He is ours, the child / ... / We only, never call him fatherless / Who has God and his mother" (IX, 409–15). This is a striking instance of the "absent,"

---

[17] According to Mermin, "The heroine's empowerment is enacted mainly through decisive revisions of the chivalric quest and rescue story which had structured Barrett Browning's imagination since childhood and which she uses here for the first time in a long poem. Much of the plot consists of thwarting Romney's grim determination to be a rescuing knight ..." (187).

"disappearing," or "invisible" father motif that is so characteristic of Victorian popular culture.[18] It is misleading, however, to see this same-sex relationship with Marian, as Aurora's ultimate personal goal in life, as some critics do,[19] virtually ignoring the elaborate, extended conclusion to the poem. In fact, after formally refusing Romney's renewed proposal and acknowledging the fact that (as perceptive readers already know) Romney and Aurora share a mutual attraction and are fated to wed, Marian more or less drops out of the story, and we assume that she will continue to find complete personal fulfillment in her role as mother. Barrett Browning is able to justify this twist of plot by associating Marian with the Virgin Mary: a kind of English Protestant Mariology is implied here,[20] in tandem with the psychological implications of her tragic sex life. It is not possible to think of Marian as a potential sexual partner for anyone, and the representation of her close relationship with Aurora makes it possible for Barrett Browning, in terms of the *Bildungsroman/Kunstlerroman* narrative, to identify with both characters: each represents different aspects of the author. Aurora is artist and wife; Marian is mother and, physically, she resembles Barrett Browning. If there was a tension between the demands of poetry and motherhood in the author's life,[21] that tension was not necessarily represented in this "autobiographical" work; rather it was avoided by dividing the two roles: Marian is the mother; Aurora is the poet–lover. There is curiously little tension between Romney's relationship with Marian and that with Aurora. Instead, both of the major female characters can be contrasted with the antagonist Lady Waldemar. Marian is not Aurora's rival – but the Lady works against, and is contrasted with, both of them in their morally legitimized relationships with Romney. Because there is finally no "competition," no sexual jealousy between the heroic Aurora and the saintly Marian–Madonna, because of their unquestioned compatibility and mutual respect and "sisterly" love, the reader feels the emotional force of a "triangle" involving Lady Waldemar, Aurora/

---

[18] In Victorian popular literature and culture, the position of the father "vis-à-vis the Victorian family was increasingly ambivalent and even antagonistic," as Claudia Nelson puts it in *Invisible Men*, her study of Victorian periodicals (1). Nelson also explores the phenomenon of the "maternal father" – in opposition to the stereotypical "cold, detached, sexually irresponsible, or brutal" father – in her essay "Deconstructing the Paterfamilias," and this version of the ideal Victorian father's commitment to domesticity can be compared to John Tosh's findings in *A Man's Place*.

[19] For example, the final chapter of Angela Leighton's book, "'Come with me, sweetest sister'": The Poet's Last Quest," ignores the massive conclusion of the poem and suggests that Aurora's relationship with Marian represents fulfillment of Aurora's quest in what is, in the words of Nina Auerbach, a "feminist hymn." Leighton also describes Aurora's father as "not only lost but irrelevant" (156).

[20] According to Isobel Armstrong, "Marian takes on not only the attributes of Mary as mother with child but also the attributes of Christ, who is through her persistently gendered as a woman" (369).

[21] Mermin comments that Aurora's "delight in Marian's son is like Barrett Browning's extravagant baby worship" (196).

Marian, and Romney. This complication is resolved with the revelation (to Aurora and the reader) that, from Romney's point of view, his relationship with the Lady had never been that of lover. In spite of all Aurora's misapprehensions, Romney was never seduced by Lady Waldemar, and his persistent attachment to the asexual Marian is based on morality and honor and, finally, does not threaten Aurora. And even when she assumes the worst about Romney, Aurora never hints at the possibility of her romantic attraction to another man.

As the principal male character in the poem, Romney's association with male violence is a particularly important issue and is closely related to his lack of potential for conventional chivalric heroism. Little physical violence is represented in the poem, although violent acts are important thematically and are central to the plot. The rape of Marian is one of the most discussed incidents in the poem because claims for the significance of Barrett Browning's work are based largely on her willingness to deal with topics that were considered indecent and inappropriate for popular literature and culture, especially as they related to women. Barrett Browning undoubtedly was thinking of references to rape and prostitution when she predicted that her work would be controversial.[22] Although some reviews criticized the poem for these reasons, negative reactions were not as prominent as she had feared. Modern studies typically focus a great deal of attention on Barrett Browning's courage in developing a character who is involuntarily taken into a brothel and raped and who subsequently becomes pregnant and delivers an illegitimate child. Marian's status as a victim is obvious. The rape itself, however, is very vaguely described. Marian is confused, disoriented. There is no rape scene; there is no direct reference to the rapist, though Marian, in telling her story to Aurora, makes it clear that "man's violence, / Not man's seduction, made me what I am" (VI, 1225–7). Marian flees from the brothel, suffers from madness, spends time tramping on the road, finds temporary employment as a maid, and finally becomes aware of her pregnancy. (When her pregnancy is revealed, she is mercilessly driven away by the woman who has employed her.)

There is one other prominent reference to an act of male violence in the poem: the incident in which Romney is injured, blinded. But here, too, even though we know that Marian's father (probably) was responsible, the incident as described by Romney is not one that the reader can visualize clearly, and we are not even certain that Marian's father, however reprehensible he is, is guilty of intentionally wounding Romney: "I do not think ... / He turned the tilting of the beam my way,– / And if he laughed, as many say, poor wretch, / Nor he nor I supposed the hurt so deep" (IX, 556–9). Of course, we know from Marian's narrative that her father has made a habit of physically abusing his wife (who, in turn, abuses Marian), and he is portrayed as a pathetic, working-class scoundrel who has been dehumanized and brutalized by a life of ignorance, poverty, and meaningless work. He is

---

[22] In her correspondence to friends following the publication of *Aurora Leigh*, Barrett Browning expressed surprise that "more offence has not been given and taken in certain quarters." For a summary of passages such as this one, see Avery and Stott, 184–5.

representative of the masses of poor and working-class people who are associated with images of mindless violence: for example, the mob that approaches the scene of the abortive wedding between Romney and Marian "clogged the streets, they oozed into the church / In a dark slow stream, like blood" (IV, 553–4).[23]

In all of this, however, there are no vividly described scenes in which individual males fight each other, and the hero Romney is never given the opportunity to rescue either Aurora or Marian from physical harm. In a larger sense, there is very little representation of intense, individual male–male competition. Romney does not compete against rivals in his courtship of either Aurora or Marian. In contrast, there is intense female–female competition between Aurora and Lady Waldemar, and in describing the unwelcome remarks of her sadistic rival, Aurora generalizes about women's capacity for psychological conflict: "'A woman takes a housewife [container for needles] from her breast / And plucks the delicatest needle out / As 'twere a rose, and pricks you carefully / 'Neath nails, 'neath eyelids, in your nostrils'" (1045–8). This is consistent with Geary's description of female competition, which "involves subtle manipulation of social relationships" (*Origin*, 68).

The lack of physical combat between males in this "modern epic" by a woman poet is consistent with its emphasis on psychological drama. To a large extent, this drama is dependent on Aurora's lack of perception, her slowness in employing "theory of mind." In spite of her "artistic sensitivity," she misunderstands Romney's intentions and is not conscious of her own deepest feelings for him until their mutual love is spontaneously revealed near the end. Aurora surely is meant to be a "reliable narrator," but, amazingly, Romney arrives for his visit near the beginning of the eighth book, and she does not discern the fact of his physical blindness until midway through the ninth book. It is only at this point, shortly after Marian retires from the scene, that Aurora discovers, through a comment Romney makes about not being able to see the stars, that he is physically blind. Incredibly, Aurora has not before this point detected signs of his blindness, not while conducting the lengthy face-to-face conversation with him and observing the dramatic exchange between him and Marian, although the alert reader has picked up numerous hints from the dialogue and from Romney's curious turning to the side at a time when it was appropriate to take her hand. Romney had sent news of his blindness in a letter that never reached Aurora, and he had assumed that she knew the truth. Now, under the shock of this discovery, Aurora begins to cry, and the wave of pity she feels for Romney triggers her passionate recognition of the longstanding love for him that she had repressed until this moment. It is interesting that Aurora's emotional reaction to the revelation of Romney's blindness is required to release her true feelings. Up to this point, Romney had taken the lead in humbling himself utterly before Aurora's greater understanding of human life and spirit, apologizing

---

[23] Some critics have condemned Barrett Browning for unsympathetic representation of the urban poor, as in such "mob scenes." For example, Cora Kaplan refers to her "vicious picture of the rural and urban poor" (11).

repeatedly for his former intellectual arrogance and lack of appreciation of Aurora's goodness and artistic insight. Now it is Aurora's turn to confess her own former shortcomings, her pride in denying Love and her womanhood by refusing to acknowledge her hidden feelings for Romney.

The acceptance of Romney's (and her own) love is dependent on her realization that love transcends art and vocation. However, despite the extraordinary resourcefulness and courage that has enabled a blind Romney to leave England and seek out Aurora and Marian in their Italian home, he must now literally depend on Aurora to see for him, performing a role that is symbolically suggested by her goddess namesake. One problem for Barrett Browning is that she wants to describe a Romney who will be dependent on an Aurora who can see and write, but he must retain a fundamentally heroic (if not chivalric) stature at the end, even if he is to be a kind of male muse for Aurora. Masculinity is in fact idealized, as suggested earlier, in a kind of gender reversal. Love is masculine, and in a sense Romney does "rescue" a woman poet who needs him and his love more than she needs anything else (including poetry). In line with traditional masculinities, however, Romney is active, not passive, in seeking out the two women and doing what he believes to be right according to his religious and moral convictions, thereby resolving the dilemmas faced by the three chief characters in the poem's plot and simultaneously fulfilling the "patriarchal" prophecy of Aurora's father.

As noted in Chapter 1, Barrett Browning has sometimes been discussed in connection with the Spasmodic school of Victorian poetry, named by hostile contemporary critics and characterized by (sometimes morbid) psychological intensity, passionate subjectivity, and an emphasis on the physical body. In an 1857 *Blackwood's* review, William Aytoun referred to *Aurora Leigh* as "fantastic, unnatural, exaggerated" (32). Recent studies of the Spasmodics emphasize the fact that they (as male poets) and the characters they created were often considered effeminate because of their emphasis on physicality and longing. For example in 1853 Kingsley published an essay in which he noted the negative influence of Shelley, whom he considered extravagant, sentimental, and "womanish" (571). As Kirstie Blair points out, Barrett Browning, in contrast to Sydney Dobell, Alexander Smith, and the other male Spasmodic poets, was "able to envisage a successful poetic hero ... who manages to write an epic poem and, in the end, conduct a passionate relationship without going mad, dying, or resorting to violence" (486). In Chapter 1, I briefly discuss Smith's poem "A Life-Drama" and problems with its very loose construction of intensely felt lyric moments tenuously held together by a flimsy romantic plot based on the life and literary career of the poet hero. It is instructive to compare Barrett Browning's work: it is as though she systematically avoided the glaring weaknesses of "A Life-Drama." The most important device can be described as novelization. As noted earlier, she thought of *Aurora Leigh* as a "novel in verse." In this way she legitimized the character development and authorial commentary associated with the Victorian novel as she constructed the life story of her heroine, with her literary career and romance.

At the same time, Aurora's identity as a *poet* invited her to incorporate intensely emotional lyrical passages that would have been awkward in prose, and the reader is expected to read the work in a manner that is more conducive not only to emotional experience but spiritual insight as well. Nevertheless, autobiographicality and the *Bildungsroman/Kunstlerroman* plot is shared by both works. Even plot details, such as Romney's confronting her with her book and poetry manuscript, have their parallels in Smith's poem, although Romney, unlike the adoring Lady in "A Life-Drama," does not appreciate Aurora's poetry, and the gender reversal, when considered in this light, is striking. What critics have not noticed, however, is that the extravagantly emotional ending of *Aurora Leigh* contrasts dramatically with that of Smith's Spasmodic epic. It is true that Walter and Violet are apparently together in love at the end, and in fact they together look out at a sunset, followed by the emerging stars, but there is nothing like the transport experienced by Aurora and Romney. Instead, Walter, who throughout the poem has responded with poetic enthusiasm to the natural imagery of the night sky, now contrasts that image with the greater "warmth" of romantic love and, as mentioned in Chapter 1, tells Violet that "A star's a cold thing to a human heart, / And love is better than their radiance" (160). Clearly Walter, whose poetic ambitions have been fulfilled by the enormous success of his published book, is deliberately downplaying both his poetic vocation and the central idea of poetry with its vision of "the infinite" growing out of a direct and powerful feeling for the natural world, and choosing the love of Violet as superior to all that. In spite of the gender reversal, Aurora is in a position analogous to Walter's at the end of her poem – her own great poem has established her poetic reputation and she is joined with the man she loves – but, far from choosing between love and poetry, she embraces both as she looks up into the night sky with her lover, who, inspired by her, has himself acquired a love of poetry and, beyond that, a poetic sensibility that allows him to miraculously "see" the spectacular night sky through the physical sight of his lover.

It might be generalized that in the context of Victorian culture she as a woman was more comfortable in dealing with powerful emotions and passionate surrender. Like other critics, Herbert F. Tucker contrasts her favorably with the male Spasmodics: unlike them, she had conscious social and political aims (including women rights) and appealed to more sophisticated poetic traditions ("Glandular Omnism," 443). However, I want to point out a distinctive kind of poetic imagery prominent in *Aurora Leigh* that guides the reader toward the conclusion of the poem and prepares the way for the union of the lovers at the end in a particular kind of poetic and spiritual vision, one fully shard by a blind Romney, who ironically has become a "poetic" hero in his own right. In Chapter 1 I refer to Glen Love, Nancy Easterlin, and others who in the tradition of "ecocriticism" emphasize the importance of the natural environment as portrayed in literature. Barrett Browning stresses Aurora's sense that she belongs to her Italian homeland in her memories of the Italian landscape in contrast with that of the English, as she negotiates her difficult transition to an English identity. In the beginning Aurora describes England in this way: "The hills are crumpled plains – the plains, parterres – / The trees, round, wooly, ready to be clipped; / And if you seek for any wilderness / You find, at best, a park. A nature tamed / And grown domestic like a barn-door fowl" (631–5). Later, her imagery of the English countryside is more positive, but nothing like that which characterizes her memories of home, and when she decides to return, she cries, "And now I come, my Italy, / My own hills! Are you 'ware of me, my hills, / How I burn toward you?" (V, 1266–8).

In the end she fails to recover her childhood "feeling" for her old home, but she is establishing a revitalized Italian identity in the context of her new Florentine home by the time Romney arrives. There is a more generalized pattern of landscape imagery, however, associated with Aurora's search for a transcendental truth in her poetic journey. Most importantly, views of – or from – physical heights elicit feelings of euphoria.[24] Specifically, this imagery is connected with Barrett Browning's spiritual quest. As Alexandra Wörn points out, "Barrett Browning sees poetry as a bridge between Paradise – the perfect state and perfect union with God – and the fallen world, bringing the reader back to a mindfulness of God" (243). Closely related to these images are those of birds, in the tradition of Shelley's "Skylark," that soar toward the sky. Both sets of images are related to the Swedenborgian correspondences discussed earlier.

These critical judgments are suggestive of Barrett Browning's relative success in negotiating the relationship between certain aspects of human nature and culture in her poetry. While insisting on Aurora's cognitive and aesthetic abilities, she allows her to retain a range of "natural," powerful emotions apparently independent of social constructions. However, Barrett Browning as female poet consciously and deliberately creates a female poet whose relationship with her "true love" at the end reverses gender roles, when Romney, as discussed, becomes a kind of male muse, fulfilling his role of male hero by furnishing emotional support to his dominant female lover – dominant not just in terms of her aesthetic position as a successful and influential creative artist but also in terms of her status as a successful professional – in the context of not socialism but political economy. Lana L. Dalley notes that Barrett Browning "presents an image of a woman who is, ultimately, both worker and wife" (531).[25] Barrett Browning boldly challenges Victorian gender categories in her representation of Aurora's status as professional female author and wife with her male muse. At the same time, however, as already implied, Aurora's life history is quite successful in terms of our most basic model of human nature. Not only have her cognitive abilities been confirmed, along with her success in earning money to supply basic human needs. She will marry the man she loves – who is extraordinarily well suited to her in every way. Not emphasized but nevertheless part of the situation at the end are the financial resources that Romney will bring to their marriage, good prospects for children in the future, and, in spite of the unorthodox positions of both husband and wife, the continuance of kin relations in a way that helps to heal the factionalism within the

---

[24] Among the studies dealing with this phenomenon are George Lakoff and Mark Johnson, *Metaphors We Live By Metaphor* (1980) and Jay Appleton, *The Symbolism of Habitat: An Interpretation of Landscape in the Arts* (1990).

[25] This is another context in which Barrett Browning opposed the views of Kingsley, whose work was associated with Christian socialism as well as traditional "manliness."

Leigh family. In spite of complications such as those related to the construction of Romney's distinctive masculinity and the marginalization of Marian Erle and her child, a meaningful connection between past and present in a larger narrative of humanity is implicit in the ending.

I have discussed the plot of *Aurora Leigh* in some detail because the poem is indeed a generic hybrid, as Dentith describes it, "moving in and out of the idioms which it deploys" (97). Tucker, who refers to the poem as Barrett Browning's masterpiece, stresses its Spasmodic qualities: "Bravely and cannily exploiting spasmody for the platform it offered women's poetry, she harnessed its authentic power to an epic design" (*Epic*, 378). The poem is a verse–novel, a *Kunstlerroman*, a modern epic, and a vehicle for protofeminist social criticism. Of particular interest in a Darwinist reading, the poet and her protagonist Aurora, in her "Spasmodic" way, emphasize a body–mind continuum and the emotional engagement of the narrator – "[T]he rhythmic turbulence / Of blood and brain swept outward upon words" (I, 897–8) – and the narrative concludes with a strong affirmation of passionate sexual love and in fact a continuity of group identity. As pointed out, in spite of the social criticism implicit in Aurora's ambivalent attitudes toward her English "home," her loyalty to her "international" father and his legacy – from his humanistic learning to his parting message of love – is strongly affirmed, and the heroic, loving cousin with whom she unites at the end is dedicated, like her, not to radical ideology of social reform but to a poetic vision of religious idealism that assumes fundamental, universal principles, affirmed by God.

Chapter 4
Clough's Ambivalent Victorian Manhood

When Arthur Hugh Clough went to Rome in April1849, at the age of 30, to witness and participate in the Roman Republic, which Mazzini and Garibaldi had proclaimed about two months earlier, he was a successful poet who had recently published two volumes, *The Bothie of Toper-na-Fuosich* (1848) and *Ambarvalia* (1849). However, as biographers and critics have pointed out over the years, he was also a man with uncertain, conflicted ideas and attitudes concerning not only art and politics but his own life and vocation. His status as Headmaster Thomas Arnold's favorite pupil at Rugby; his religious doubts and involvement in the Tractarian controversy during his undergraduate years at Balliol College, Oxford, and his disappointing second-class degree there; his dramatic resignation of his Oriel Fellowship at Oxford in 1848 because he decided that he could not in good conscience subscribe, as required, to the Thirty-nine Articles of the Church of England: all of these striking details of his life are well known, as is his important, complex relationship with Matthew Arnold, who celebrated his brilliant friend as a sensitive but tragically self-conscious and indecisive poet in the elegy "Thyrsis" (1866).[1] All major studies of *Amours de Voyage*, which was written primarily during Clough's three-month stay in Rome but revised through the years and not published until 1858, point to obvious connections between Clough and his narrator–protagonist Claude, though it is important to remember that Claude is a fictional character whose point of view does not necessarily correspond to that of the author. After resigning his fellowship in 1848, Clough, who identified himself as a republican and was attracted to political radicalism, journeyed to Paris to witness the revolution of that year firsthand prior to his trip to Italy, and his ambivalence about both the Italians and the French, who invade to bring down the short-lived Italian republic, is reflected in *Amours*, along with his doubts about religion, poetry, and romantic love. But admirers of this work also point to an integrity and intellectual honesty in both Clough and Claude that does not submit to reductive cynicism.

Clough's relationship with Claude, then, is in some ways analogous to that between Barrett Browning and Aurora. Aurora is a young woman poet who has progressive social and religious attitudes that the reader is invited to associate with

---

[1] Through the years several Clough scholars have claimed that Arnold's poem misrepresents his friend. For example, Kenny in his biography of Clough argues that "there is a grave distortion in his comparison of Clough's search for truth and the wanderings of the Scholar Gypsy" (287).

those of her creator and an ambivalence toward the English and her father's family but who is not fully connected with her Italian homeland, either. She has a certain rootlessness that enables her to define her own distinctive artist's vision. Claude is an English *tourist* in Europe who insists on his independent, subjective view of things, too. Like Barrett Browning, Clough is an experimental poet who "novelizes" his poem and emphasizes a problematic but promising love relationship; at the same time, his hexameter verse form is suggestive of the ancient epic tradition that held interest for Barrett Browning (though in this context it is ironic, as I discuss below). However, Aurora not only finds fulfillment in her poetic vocation but finds true love in Romney, whereas Claude exhibits no clear sense of vocation, does not comfortably adjust to the confused world around him, and is unable to unite with his potential wife Mary.

The final version of *Amours,* which Clough first published in *The Atlantic Monthly* in 1858 and which appeared in volume form only in 1862, after his death the previous year, consists of five cantos. The poem is primarily in the form of letters, most of them written by the English traveler Claude to his friend Eustace, with some written by other correspondents, as described below, and introductory and concluding elegiac lyrics for each canto in a narrative voice (that often seems to merge with Claude's voice). Clough himself referred to the structure as a "five-act epistolary tragic-comedy, or comi-tragedy."[2] In the first canto, Claude tells Eustace in an oft-quoted passage that "Rome disappoints me much; I hardly as yet understand, but /*Rubbish*/ seems the word that most exactly would suit it" (I, 19–20); however, "Rome is better than London because it is other than London" (26). The setting is established, with references to the French invasion of the Roman Republic, and Claude refers to the English Trevellyn family, with whom he has been spending time, and especially the daughter Mary, who is identified as his special friend. His nonchalant, satirical attitude toward Rome, marked by its pretensions of grandeur and its pagan past – "No, great Dome of Agrippa, thou art not Christian, canst not, / Strip and replaster and daub and do what they will with thee, be so" (I, 156–7) – is consistent with his attitude toward the Trevellyns – "Middle-class people these, bankers very likely, not wholly / Pure of the taint of the shop" (I, 125–6), "Yet they are fairly descended, they give you to know, well connected" (I, 129). Occasionally interspersed with letters from Claude to Eustace are ones from Georgina Trevellyn to her friend Louisa, along with occasional notes by Mary Trevellyn herself, sometimes incorporating her impressions of Claude. Especially significant is a key word used by Claude when he describes growing intimacy with the Trevellyns, and especially Mary: "Well I know after all it is only juxtaposition; / Juxtaposition is short; and what is juxtaposition?" (I, 229–30). The randomness of his acquaintance with her in the opening chapter of this "romance" is stressed in the context of an historical setting that is imposing but marked by hollow and misleading claims.

---

[2] See Clough's letter of 10 April 1858 to F. J. Child, in *The Poems and Prose Remains*, ed. Blanche Clough, I:232.

The position of Claude as an intellectually engaged yet emotionally detached narrator is more complex than might at first appear. Anthony Kenny describes him, accurately enough, as a "supercilious Oxford don on the Grand Tour," writing "to his clerical friend Eustace in London" (164). A somewhat haughty, sarcastic young Englishman reporting on the political and military situation and mocking tourists' perception of "the glories of ancient Rome" in this situation might be of limited interest, but Claude is more than that. Even if we set aside the overly simplistic and misleading scheme of identifying Claude with Clough himself, it is reasonable to see him as representing to some degree the relatively sophisticated intellectual and cultural background of his creator. Unlike Clough, whose heartfelt commitment to republican values inspired his visit to Rome and who had the connections that enabled him to visit Mazzini himself,[3] Claude is much more the casual tourist. Nevertheless, though Claude does not elaborate on his educational background or position, he, like Clough in his extensive correspondence to his old friends, speaks with the voice of one who assumes a knowledge of Roman history in the context of a broad awareness of Western civilization like that attained by one who has been deeply engaged in the "literary" curriculum at Rugby and then Balliol College, Oxford. His current status as a tourist frees him not only from the intensely political conflict swirling around him but allows him to identify himself in a larger sense as a disengaged, uncommitted man who nevertheless is acquainted with associations powerfully attached to cultural, national, and religious ideals. His intellectual sensibility and large perspective allow him to identify with the plight of the Romans while he understands that their political cause is probably hopeless, and his own sympathy for republican ideals, which he dares not allow to dominate his mind, is apparently associated with a past commitment to religious idealism in his own life that gives him some insight into the classical pagan, then Christian history of this ancient capital of the Western world. The introductory elegiac stanza or prologue conjures up the glories of the Roman past.

> *Over the great windy waters, and over the clear crested summits,*
>   *Unto the sun and the sky, and unto the perfecter earth,*
> *Come, let us go,-- to a land wherein gods of the old time wandered,*
>   *Where every breath even now changes to ether divine,*
> *Come, let us go; though withal a voice whisper, "The world that we live in,*
>   *Whithersoever we turn, still is the same narrow crib;*
> *'Tis but to prove limitation, and measure a cord, that we travel;*
>   *Let who would 'scape and be free go to his chamber and think;*
> *'Tis but to change idle fancies for memories wilfully falser;*
>   *'Tis but to go and have been." –Come, little bark, let us go!* (I, 1–10)

The point of view here is one that the author Clough assumes is compatible with that of his implied audience. The reader understands that Claude is speaking to a

---

[3] According to Kenny, Clough, after arriving in Rome, "demanded an audience with Mazzini" and the meeting took place on 22 April 1849. See *Arthur Hough Clough*, 156–7.

friend, a fellow English intellectual who places Rome into historical perspective and can smile at descriptions of its "rubbishy" contemporary appearance while appreciating its mythic past and understanding an attempt to reconcile Rome's Catholic Church traditions with contemporary Protestant Christian traditions and moral values in the broadest sense. But the problem of Western civilization and its values is still more complex and far-reaching. Late in Canto I, Claude finds himself admiring the statues of the Horse Tamers at the Monte Cavallo fountain:

> Ye, too, marvelous Twain, that erect on the Monte Cavallo
> Stand by your rearing steeds in the grace of your motionless movement,
> Stand with your upstretched arms and tranquil regardant faces,
> Stand as instinct with life in the might of immutable manhood,--
> O ye mighty and strange, ye ancient divine ones of Hellas,
> Are ye Christian too? To convert and redeem and renew you,
> Will the brief form have sufficed that a Pope has set up on the apex
> Of the Egyptian stone that o'ertops you, the Christian symbol?
> And ye, silent, supreme in serene and victorious marble,
> Ye that encircle the walls of the stately Vatican Chambers,
> Juno and Ceres, Minerva, Apollo, the Muses and Bacchus,
> Ye unto whom far and near come posting the Christian pilgrims,
> Ye that are ranged in the halls of the mystic Christian pontiff,
> Are ye also baptized? are ye of the Kingdom of Heaven?
> Utter, O some one, the word that shall reconcile Ancient and Modern;
> Am I to turn me for this unto thee, great Chapel of Sixtus? (190–205)

Claude's joking manner concedes an implicit admiration for this monumental sculpture: it may not reconcile ancient and modern, Christian and pagan, but it inspires the educated observer to think of the classical Egyptian, Greek, and Roman traditions that anticipate the present age in all its complexity and change, as curious English Protestant and agnostic tourists traverse the historical landmarks and troops representing a European nation closely associated with progressive, republican values, now under the rule of Louis-Napoleon as an elected President, prepare to destroy a fledgling Roman republic and reinstate the authority of the Pope. For Claude, a sense of awe and wonder can still penetrate the disgust and boredom he experiences as he wanders though the "rubbishy" tourist attractions. It is significant that Claude sees the faces of the horse tamers "as instinct with life in the might of immutable manhood." Throughout the poem, references to manhood are associated with strength and bravery, especially in connection with battle and the protection of women. Unquestioned, dedicated manhood, like religious faith, belongs to the past: anxiety over this possibility is implicit in much of Claude's commentary and is central to the issue of masculinity in the poem.

On a more obvious level, as Clough introduces the courtship theme in Canto I, Claude's impressions of Mary and the Trevellyn family are from the beginning placed in the context of his musings about art, religion, and sexuality. Following a short letter in which Georgina Trevellyn tells her friend Louisa how she and her family are "delighted of course with St. Peter's /... / Rome is a wonderful place" (I, 54–6), Claude writes Eustace that "the Christian faith, as I, at least, understood it, / Is not here, O Rome, in any of these thy churches, / Is not here, but in Freiberg, or Rheims, or Westminster Abbey" (I, 70–73). In St. Peter's and some of the churches, he finds the works of the "sixteenth-century Masters," representative of a "positive, calm, Stoic-Epicurean acceptance" that is "Overlaid of course with infinite gauds and gewgaws, / ... / Forced on maturer years as the serious one thing needful, / By the barbarian will of the rigid and ignorant Spaniard" (I, 76–82). Claude then recalls the legacy of Luther, who may have been "unwise" and even "foolish," but "oh, great God, what call you Ignatius?" Claude's contempt for the Counter-Reformation is expressed in emotional outbursts: "these vile, tyrannous Spaniards," / These are here still, – how long, o ye heavens, in the country of Dante?" (I, 103–6). Claude goes on to refer to "emasculate pupils and gimcrack churches of Gesu, / Pseudo-learning and lies" (I, 105–10). The pseudo-learning of the Spanish religious orders is associated with their lack of masculinity.

The hypercritical Claude, while consistent in his attack on the Jesuits and Counter-Reformation, "Spanish" Catholicism in general as he tours the ancient city, rocks back and forth in more playful, tentative criticism of English bourgeois tourists like the Trevellyns – "Ah, what a shame indeed to abuse these most worthy people! / Ah, what a sin to have sneered at their innocent rustic pretensions!" (I, 135–6) – and he compares himself to Shakespeare's evil character Iago, who "can be nothing at all, if not critical wholly" (148). He finds it "pleasant, I own it, to be in their company," especially pleasant to "abide in the feminine presence," and he goes on to speculate about the desire among "some poor soft souls" for "a happy serene coexistence," a "necessity simple, / Meat and drink and life, and music filling with sweetness, / Thrilling with melody sweet, with harmonies strange overwhelming, / All the long-silent strings of an awkward meaningless fabric" (172–6). Most strikingly, he detaches the desire for children from the "dream of romance" and "fever of flushed adolescence" and suggests a kind of "avuncular" fulfillment:

> Nephews and nieces! alas, for as yet I have none, and moreover
> Mothers are jealous, I fear me, too often, too rightfully: fathers
> Think they have title exclusive to spoiling their own little darlings;
> And by the law of the land, in despite of Malthusian doctrine,
> No sort of proper provision is made for that most patriotic,
> Most meritorious subject, the childless and bachelor uncle. (184–9)

Clough assumes that his (English) readers will be willing to follow him as he dissociates Claude from allegiances to the values of "narrow" English middle-class families while retaining a certain interest in his countrymen, though his protagonist/narrator is something of a snob. Claude's elitism is probably meant to be seen as a flaw in his character,[4] but Claude assumes an English identity as he

---

[4] As Kenny points out in his biography, "Claude has several characteristics which set him apart from Clough. He is superior to him in wealth and status ... The real Clough was more likely to hear than to make sneering remarks about those whose family are 'not wholly pure of the taint of the shop'" (173).

maintains an important friendship with Eustace, back home in England, with whom he shares his candid thoughts. Like Barrett Browning, Clough is writing a kind of novel in verse, but his is of a different sort. The epistolary mode hearkens back to the early eighteenth-century novels – for example, Samuel Richardson's classic *Pamela; or, Virtue Rewarded* (1740) – that had established the genre of the novel, although in *Amours* the dialogue or "conversation" is one-sided, and the reader sees only Claude's letters (in an early version of the manuscript Clough included a few of Eustaces's letters but later decided to omit them).[5] Claude's letters in effect constitute a series of monologues, though the "audience" of a fellow young intellectual Englishman is important in the same way an implied audience is often significant in Robert Browning's dramatic monologues.

As for the verse form itself, Clough could assume that his hexameters would be associated in many readers' minds with the ancient Homeric epic tradition. In 1860–61, Matthew Arnold, as Professor of Poetry at Oxford, would give a series of three lectures entitled "On Translating Homer," and, influenced by Clough's efforts, he had been experimenting with English hexameters, though he himself undertook no large translation projects.[6] Clough had already used hexameters in *The Bothie*, where he makes heroic/classical allusions especially clear for satiric effect. It is ironic that his use of the hexameter in *Amours*, while criticized by many as awkward (some lines cannot be scanned in the proper way), has also been praised for accommodating an idiomatic, conversational English style.[7] Clough adapts seemingly archaic literary forms to achieve his artistic aims, which are very much those of a contemporary young Englishman who compulsively questions traditional cultures and ways of life, but, of course, it is the novelistic plot that appeals to readers' interest in Claude's destiny. Significantly, however, it is difficult to "visualize" Claude and Mary together. In stark contrast with *Aurora Leigh,* there are no dramatized scenes in which Claude interacts with Mary and her family, no conversations or direct exchanges of letters – and the emotional responses of both Claude and Mary are indirectly expressed, implicit in the texts of their letters.

Mary Trevellyn, in a postscript to a letter by Georgina (the final letter in Canto I), refers to Claude: "He is a what people call, I suppose, a superior man, and / Certainly seems so to me, but I think he is frightfully selfish" (I, 273–4); indeed, Clough has encouraged the reader to see Claude as a somewhat egotistical

---

[5] See Scott's account of Clough's revisions of *Amours* in the introduction to his edition (3–16). He notes that "Eustace's common-sensical replies" to Claude "had tended to deflate the reader's view of Claude" (10).

[6] See Machann, *Matthew Arnold*, 47.

[7] There are significant differences between Clough's English hexameters and those of his classical sources. In Latin and Greek, a foot of verse is defined by the length of the syllables occurring in it, but in English verse, it is defined by the position of the stress or accent. Furthermore, Clough often uses a trochee in places where classical poets used a spondee. Nevertheless, Biswas, for example, praises the "superbly original artistry" of Clough's hexameters (321).

protagonist, but the elegiac postscript, implicitly from Claude's point of view, returns to a philosophical perspective:

> *Alba, thou findest me still, and, Alba, thou findest me ever,*
> *   Now from the Capitol steps, now over Titus's Arch,*
> *Here from the large grassy spaces that spread from the Lateran portal,*
> *   Towering o'er aqueduct lines lost in perspective between,*
> *Or from a Vatican window, or bridge, or the high Coliseum,*
> *   Clear by the garlanded line cut of the Flavian ring.*
> *Beautiful can I not call thee, and yet thou hast power to o'ermaster,*
> *   Power of mere beauty; in dreams, Alba, thou hauntest me still.*
> *Is it religion, I ask me, or is it a vain superstition,*
> *   Slavery abject and gross, service, too feeble, or truth?*
> *Is it an idol I bow to, or is it a god that I worship,*
> *   Do I sink back on the old, or do I soar from the mean?*
> *So through the city I wander and question, unsatisfied ever,*
> *   Reverent so I accept, doubtful because I revere.* (275–7)

Claude may be "frightfully selfish" in his social interactions, but he is also sincerely, emotionally engaged, in a search for meaning in life. The idea of Rome, the myth of Rome, is of vital interest to him, and as he wanders among the monuments that define the historical city in its various phases, he continually looks up to the Alban Mount, visible in the distance, mythic site of the civilization that predated Rome, associated with the god Jupiter, with the city of Alba Longa, and the story of Romulus (and with Aeneas, in Virgil's epic). If Rome conjures up images of ancient origins of civilization and religion, Alba reaches back further. In the context of historical and mythic change, where is the foundational truth? The *reverent* thinker in his search for that truth cannot reject an honest doubt. The context of Claude's classical, literary education here on a surface level makes his affinity for this setting realistic, but it also channels his emotionally charged search for ways to organize his most basic desires in terms of the behavioral systems discussed in previous chapters. Claude's personal thoughts about mating, parenting, and social relations directly related to the romantic plot of the narrative are not simply placed into a "charming" physical setting; on the contrary, they are juxtaposed (to adapt Clough's use of that key word) to the *universal* themes associated with human nature. Claude's feelings about his masculinity are central to this pattern of meaning in the narrative, and his observations and speculations about *chivalric* manhood as related to battle are especially emphasized in Canto II.

In the first letter of the second canto, Claude evaluates the political situation as the French troops prepare to invade, and although he sets himself apart from visions of republican idealism associated with the French Revolution, "I nevertheless, let me say it, / Could in my soul of souls this day with the Gaul at the gates, shed / One true tear for thee, thou poor little Roman republic." As for France, "it is foully done" and of course "my stupid old England," which supposedly supports the self-determination of nations, will not interfere, though the English will read about the situation in the *Times* (II, 20–25). In his second letter, he meditates on the idea of dying for "the Cause." "On the whole we are meant to look after ourselves" (37), and the idea of sacrificing one's life in order to defeat enemies seems odd. Nature intended the individual to cling to his own life. Of course, Clough is exploring very basic ideas associated with masculinity and war, and it is as though he is systematically deconstructing the chivalric values defined in Tennyson's *Idylls*. Allegiance to the Roman republic is insufficient even for most Romans: "Sweet it may be and decorous perhaps for the country to die, but / On the whole we conclude the Romans won't do it, and I shan't" (46–7). As Claude continues to report on the prospects for real battle between the French and the Romans, he repeatedly returns to the idea of placing his own life in jeopardy in defense of a cause. In letter iii, he has a waking dream of "great indignations and angers transcendental, / Dreamt of a sword at my side and a battle-horse underneath me" (61–2) but when in the following letter, he asks himself if "by some evil chance," French troops confront the English in Rome where they are gathered for safety in the Maison Serny, "Am I prepared to lay down my life for the British female?" his answer is, "Really, who knows?" (66–7). As Houghton points out in his reading of the poem, "To question the validity of dying for the women of England is bad enough. To refer to them as 'females' is to betray the whole code of the chivalric hero. This *was* shocking" (*The Poetry of Clough*, 129).

Significantly, Clough himself acknowledges the "shocking" nature of Claude's statement and has him analyze it in his own way: "One has bowed and talked, till, little by little,' / All the natural heat has escaped of the chivalrous spirit. / Oh, one conformed, of course; but one doesn't die for good manners, / Stab or shoot, or be shot, by way of graceful attention" (67–70). Claude explains that he would be much more likely to die "on the barricades" fighting "for the vapour of Italy's freedom" and that he can imagine himself protecting a child in the street, but he does not feel the "vocation" to protect "the full-blown lady-- / Somehow, Eustace, alas, I have not felt the vocation. / Yet these people of course will expect, as of course, my protection" (73–7). That is, the Trevellyn family naturally assumes that Claude will join George Vernon, who is engaged to Mary's sister Christina, in defending members of the family, especially the women, if the unstable situation generated by the French attack leads to physical danger. At issue here is Claude's responsibility as a young Englishman, intensified by his socialization with the Trevellyns, and Mary in particular. Claude understands very well that it is "unchivalric" for him to question and analyze what should be a spontaneous, emotionally charged commitment to do his duty. He also believes his reader Eustace expects that "When the time comes you will be ready" (81), but how can he assume that will be the case: "What I cannot feel now, am I to suppose that I shall feel? / Am I not free to attend for the ripe and indubious instinct?" (81–4). In describing his dilemma, Claude combines ideas of supposed social and moral duty along with that of strong feeling or instinct in a way that would disturb not only his correspondent Eustace but some of Clough's friends when he showed them his poem in manuscript. I take up this issue in more detail later in the chapter, but here I want to emphasize the point that Claude is very much aware of his apparent "cowardice" and that what he is trying to do here and in the rest of the poem is to explain and justify his sincere but obviously inappropriate thoughts and feelings, not just to his sympathetically inclined friend Eustace but to himself: "Is it the calling of man to surrender his knowledge and insight, / For the mere venture of what may, perhaps, be the virtuous action?" (86–7). He is willing to consider the possibility that "all this" is "but a weak and ignoble refining, / Wholly unworthy the head or the heart of Your Own Correspondent?" (92–3).

The military crisis in Rome is real, but the detached and analytical Claude experiences it at a physical and psychological distance. As before, Claude rejects the fashion of Romantic subjectivity in exchange for dramatized narration and satire, but his ironic stance becomes increasingly evident. In letter v of Canto II, he enters the Café Nuovo, his Murray guidebook[8] in hand, "as usual," thinking about his projected view of the Campidoglio Marbles (in the museum on the Capitoline Hill), but because of the impending battle on that day (April 30), as the French prepare for their first attack on the city, there is no milk for his coffee: "Yes, we are fighting at last, it appears" (95). Note the ironic use of "we." The first French attack is repulsed by the Romans, as Claude watches the distant puffs of cannon smoke from the Pincian Hill, where he stands "with lots of English, Germans, Americans, French, – the Frenchmen, too are protected" (113–14). Claude is of course pleased that the French have been turned back: "Victory! Victory! Victory! – Ah, but it is, believe me, / Easier, easier far, to intone the chant of the martyr / Than to indite any paean of any victory. Death may / Sometimes be noble; but life at the best will appear an illusion" (148–50). He conjures up the image of death suffered in battle; the "smoke of the sacrifice rises to heaven, / Of a sweet savour, no doubt, to Somebody; but on the altar, / Lo there is nothing remaining but ashes and dirt and ill odour" (152–5). The sacrifice of the fallen may be sweet to some (imagined?) Deity – "Somebody" – but this conjectural "spiritual" significance is contrasted with the physical realty.

In the following letter, Claude describes his having seen "a man killed! An experience that, among others! / Yes I suppose I have; although I can hardly be certain, / And in a court of justice could never declare I had seen it" (162–4). A priest, attempting to "fly / to the Neapolitan army" (190–91) was apparently killed by a hostile Roman mob. Claude, observing the incident, while walking back to his residence from a tour of St. Peter's, his Murray guidebook under his arm, is not quite sure about what he has witnessed, but he sees swords "smiting, / hewing chopping" (184–5), takes note of the blood, and sees "through the legs of the people the legs of a body" (197). According to rumor, as many as five priests may have been thus killed by the Roman crowds, but Claude is thankful that the government is strong enough to stop this kind of violence.

At this point, Clough inserts a short letter by Georgina in which she makes reference to Claude's being "most useful and kind on the terrible thirtieth of April" – and not selfish after all – but wonders "what can the man be intending" and

---

[8] John Murray's *Handbook for Travellers in Central Italy* (1843).

remarks that "Mary, who might bring him to in a moment, / Lets him go on as he likes, and neither will help or dismiss him" (226–34). She also mentions (obviously unsubstantiated) rumors about Garibaldi's cruelty in employing an American negro who rides behind him and kills enemies by strangling them with his *lasso*, and she is critical of the "dreadful" Mazzini (in contrast to Claude's admiring references to the "noble" Mazzini) for seizing horses throughout the city, making it difficult for her father to obtain the horses needed for their planned exit from the beleaguered city. Most significant in terms of the plot is the reference to the Trevellyns' impending journey to Florence in this letter, followed by one in which Claude makes positive statements about Mary: "It is a pleasure indeed to converse with this girl" who "can talk in a rational way" and "Speak upon subjects that really are matters of mind and of thinking, / Yet in perfection retain her simplicity" (253–6). And yet in response to Eustace's inquiry whether he is in love with her, he replies with an evasive phrase, in what is probably the most frequently quoted line from the poem: "I am in love, you say; I do not think so exactly" (263). This is the first of a series of four letters in which Claude attempts to explain his opinions about romantic love and his confused feelings for Mary. His fear of commitment is clear in statements such as, "I do not like being moved; for the will is excited; and action / Is a most dangerous thing; I tremble for something factitious, / Some malpractice of heart and illegitimate process; / We are so prone to these things with our terrible notions of duty" (270–73). Claude is afraid of being moved by an artificial or inauthentic sense of duty rather than a genuine, natural impulse. He wants to take his time and make sure that he feels the real "inspiration" of love: "Ah, let me look, let me watch, let me wait, unhurried, unprompted" (274). He is afraid that Mary "doesn't like me, Eustace, I think she never will like me. / Is it my fault, as it is my misfortune, my ways are not her ways?" (282–3). And yet, although the situation seems hopeless, "I cannot, though hopeless, determine to leave it: / She goes, therefore I go; she moves, I move not to lose her" (288–9). As Claude analyzes his desires and impulses, it is as though Clough is using the distinctive historical and cultural setting of his poem as a kind of laboratory in which to test the "natural" responses of his intellectual but fully humanized hero (or anti-hero).

 As Canto II comes to an end, the double vision of Claude is laid out for the reader. He is not in love "exactly," but he is attracted to Mary. He enjoys talking with her and admires her good sense but at the same time he feels that "she doesn't care a tittle about me!"(301). He understands that "'tisn't manly, of course, 'tisn't manly, this method of wooing; / 'Tisn't the way very likely to win. For the woman, they tell you, / Ever prefers the audacious, the willful, the vehement hero," and if she does not appreciate "knowledge," he himself does not really desire it either (290–95). And, after all, he would like for Eustace to meet this woman, who "is certainly worth your acquaintance. / You couldn't come I suppose, as far as Florence, to see her?" (315–16). In the final letter of Canto II, Georgina reports that the family is starting for Florence, and, as Claude might expect, in a P.S. by Mary and another by Georgina, Claude is compared unfavorably to George Vernon. Mary thinks Claude "agreeable, but also a little repulsive" (330) and observes that one "satisfactory marriage" in the family will do for that year. Georgina remarks that Claude and her father are great friends, though Claude "really is too *shilly-shally*, / So unlike George" (335–6). Still, she hopes that Claude and her sister will get together, and she hints that she will ask George to say something to Claude before the family leaves. The elegiac postscript to Canto II implies that indeed Claude will follow the Trevellyn party as they leave for Florence – "anon he shall follow" (343). The conventions of romantic narrative in fact demand that, after further difficulties and complications, Claude will be united in marriage with Mary, with or without the support of Mary's family and Claude's friends. Beyond nineteenth-century British and European and American narrative conventions, Hogan's outlines of universal prototypical narrative plots as discussed in Chapter 1 would lead us to expect that Claude and Mary, assuming that they are both sympathetic characters, will get together in the end. And yet this will not happen in the conclusion of Clough's remarkable narrative poem – to the amazement and disappointment of Ralph Waldo Emerson, who read the poem as it was being published in *The Atlantic Monthly*, and other friends of Clough who read the poem in manuscript, as discussed below. Because the two most prominent universal prototypes for *happiness* in narratives according to Hogan are "romantic union with one's beloved" and "the achievement of political and social power, both by an individual and by that individual's in-group (for example, his/her nation)" (121), it is no wonder that many of Clough's readers have not been happy with his conclusion.

The "dialogue of the mind with itself," in Matthew Arnold's phrase, intensifies in Canto III, as Claude, still in Rome, thinks about Mary in Florence and in the course of his lengthy discourse on the meaning of life decides that indeed he should follow the Trevellyns and offer a proposal of marriage, but, as before, his self-examination is framed in the implicit dialogue with his friend Eustace. However, an opening letter from Mary to Miss Roper, her former governess, records her reaction to Claude's odd and unexpected decision to remain in Rome and visit the "Vatican marbles" rather than accompany the family to Florence: "I really almost am offended." Claude is sometimes "repulsive," though he can be "quite unaffected, and free, and expansive, and easy" when he talks about ideas. She describes him as a "cold intellectual being" who does not "make advances." He seems to think "that women should woo him; / Yet, if a girl should do so, would be but alarmed and disgusted. / She that should love him must look for small love in return" (III, 35–7). But a bit later in the Canto is a letter in which Mary inquires of Miss Roper, still in Rome, whether she has seen Claude, indicating a continuing interest.

The counterpoint of Claude's letters to this commentary on his character becomes increasingly pronounced as he asks Eustace, "do you think that the grain would sprout in the furrow, / Did it not truly accept as its *summum* and *Ultimum bonum* / That mere common and may-be indifferent soil it is set in?" (40–43) and goes on to speculate about man's relationship to nature. Unlike the seed, man is a conscious being who can speculate about the great scheme of life, can analyze his own human nature. But Claude remembers looking into the sea as he journeyed from Marseilles to Civita Vecchia (the port for Rome) and associates this experience with the image of a statue of the sea god (merman) Triton that he later saw in the Vatican museum, remarking that "we are still in our Aqueous Ages." Clough is evoking classical mythology in a way that connects it with the new science – geology and biology. Houghton interprets this passage to mean that we are "natural beings" who still have natural instincts and we have emerged from the water "only slightly, painfully, into a higher realm of rational thought and perspective. It is much too soon for man to think of abandoning the natural life" (145). Of course it is one thing for Claude to understand – intellectually – that people are motivated by primeval urges and another to feel himself caught up in the desire to pursue Mary. In the meantime, he returns to the related "primal" topic of battle and ridicules the idea that his sympathy for the Romans could motivate him to join the fight: "Why not fight? – In the first place, I haven't so much as a musket. / In the next, if I had, I shouldn't know how I should use it. / In the third, just at present I'm studying ancient marbles. / In the fourth I consider I owe my life to my country. / In the fifth, – I forget; but four good reasons are ample" (68–72). The reader may find this playful self-satire a bit too much to take, however, when he follows it with the comment "let 'em fight, and be killed. I delight in devotion" and refers to "the glorious army of martyrs" (73–4). After this, he devotes a letter to the Biblical Tree of Life, with its "transient blossom of knowledge" at the top, "Flowering alone, and decaying,"(84), reinforcing his growing concern that human consciousness itself, apart from *living life*, is worthless: "we are still in our Aqueous Ages" (97). Claude devotes himself to knowledge, his only sure value, but his knowledge tells him that there is more to life, and he becomes increasingly aware of the limitations of a mind–body dualism. Clough implies that poetry may have the power to transcend a mind-body dualism, but he must try to do this in his own way. He may be writing in hexameters, but they are nineteenth-century English hexameters, and he is a modern English poet writing to a modern English audience.

In letter vi, returning to the idea of "juxtaposition," Claude conjures up the image of meeting a woman in a "railway-carriage, or steamer" and "*pour passer le temps*, till the tedious journey be ended, / Lay aside paper or book, to talk with the girl that is next one, / Talk of eternal ties and marriages made in heaven" (111–12) but he goes on to extend the idea of contrasting the present reality with the prospect for eternity and asks, "But for his funeral train which the bridegroom sees in the distance, / Would he so joyfully, think you, fall in with the marriage-procession?" (117–18). Here Claude is not thinking only of a transient worldly existence terminated by death but also "a freer and larger existence" beyond. Claude no doubt intends to suggest not an orthodox Christian version of the afterlife but a transcendental spiritual vision, the kind of Carlylean adaptation of German philosophy that attracted Clough's interest at this stage of his career. What stands out here is the contrast between a spirituality, which Clough surely associates with the conflicted but all-important religiosity that he and various male friends at Oxford had shared, and what he takes to be the typical attitude of women: "Ah, but the women, – God bless them! – they don't think at all about it" (130). And yet men must "eat and drink," involve themselves in the physical world, and accomplish in God's sight "our petty particular things," confident that knowledge will abide with Him if it is lost with men. "Juxtaposition is great" but the maiden:

> Hardly would thank or acknowledge the lover that sought to obtain her,
> Not as the thing he would wish, but the thing he must even put up with, –
> Hardly would tender her hand to the wooer that candidly told her
> That she is but for a space, an *ad-interim* solace and pleasure, –
> That in the end she shall yield to a perfect and absolute something,
> Which I then for myself shall behold, and not another,-- (140–45)

Claude assumes that women want to hear promises of eternal love and are shocked and revolted by sincerity and truth. This deconstruction of the romantic "religion of love" most obviously clashes not with Victorian poetry by women but rather with the paradigm of immortal love after death imagined powerfully in Dante Gabriel Rossetti's contemporary poem *The Blessed Damozel* (1850).

Claude continues to refine his ideas as he thinks about observing nature and looking beyond mere "juxtaposition" to "affinity," a term that contemporary chemists used in a technical sense to refer to the tendency for certain substances to combine with others. As he walks in a natural setting, Claude observes living things – oxen, asses, dogs, kittens, lizards, swallows, and insects (!) – and feels "Something of kindred, common, though latent vitality, greet me" (169). Claude is trying hard to break down the profound mind–body dualism that is central to his character and which, as he now is coming to realize, has prevented him from experiencing life in a natural way. "Life is beautiful, Eustace, entrancing, enchanting to look at" (176), from the streets of a city to a chamber full of pictures to the "beautiful Earth" itself – if only we could eliminate "This vile hungering impulse, this demon within us of craving" (180). Immediately after Claude reminds himself of the "vile hungering impulse" that inevitably accompanies (and enables) natural beauty at the end of letter viii, letter ix opens with a reference to "*Mild monastic faces in collegiate cloisters*," subsequently described as a "celibatarian phrase" and a "Tribute to those whom perhaps you do not believe that I can honour" (183–5) and thus in context related to the home of "celibate" fellows at Oxford that might be identified with Claude's experience and with that of Eustace (though Patrick Scott in his notes to his edition observes that Clough had been visiting cloisters of Catholic monks in Rome).[9] Then, apparently in response to Eustace's suggestion that Claude has an "obligation" to Mary, Claude declares, "I cancel, reject, disavow, and repudiate wholly / Every debt in this kind, disclaim every claim, and dishonour" (192–3) for "she held me to nothing" (197). But his exclamation "*HANG* this thinking, at

---

[9] See *Amours* (33, n. 182).

last! What good is it? oh, and what evil" (207) implicitly signals Claude's resolve to follow Mary. Then an important plot detail is disclosed in comments made by Mary and by Claude in final letters at the end of Canto III. Mary remarks to Miss Roper that George Vernon, "before we left Rome, said / Something to Mr. Claude about what they call his attentions. / Susan two nights ago for the first time heard this from Georgina. / It is *so* disagreeable and *so* annoying think of" (240–42). It appears that Miss Roper, still in Rome, has seen Claude and in a conversation has heard something of this, and Mary is unsure of how to respond but asks her friend "Not to let it appear that I know of that odious matter" (256). In Claude's version of the story for Eustace, "the foolish family Vernon / Made some uneasy remarks, as we walked to our lodging together, / As to intentions forsooth, and so forth. I was astounded" (273–4), and this explains Claude's decision to find an excuse to dally in Rome while the Trevellyn family, minus the honeymooning George and Georgina Vernon, depart for Florence. Afterwards, however, Miss Roper, at Mary's request, passes along to Claude the information that George had acted on his own, without Mary's knowledge. In Canto IV, Claude gives up his role as tourist and strikes out in an attempt to catch up with the woman he seems destined to marry.

At this point Claude is playing the role of a somewhat conventional romantic hero, an "unfortunate hero," as Clough described him.[10] He arrives in Florence but discovers that the Trevellyns have already left for Milan. He rushes to Milan but finds them gone. As we know from Mary's correspondence to Miss Roper, she has left a letter for Claude explaining that the family has left for Como. Mary's letter is lost by the *cameriere,* but Claude learns of their destination from another source and hurries to Como. Once again, they have already left and Claude has no reliable information about their destination this time. He goes to the Splügen pass in Switzerland, to the Stelvio pass in Austria, then on to Porlezza, looking in vain for the family name in one visitors' book after another. Returning to the inn at Como, he examines the book again and finds a note "*By the boat to Bellaggio.*" Indeed, Mary had written this clue to guide his search, but the family's plans had been altered and they had gone to Lugano instead. In Lugano, Mary writes a similar note, "To Lucerne, across the St Gothard." However, a frustrated Claude is still in Bellaggio (on the shore of Lake Como, across the border into Austrian Lombardy), writing Eustace about his plight. Adapting a famous passage from Shakespeare's *Julius Caesar*, he declares:

> There is a tide, at least in the *love,* affairs of mortals,
>   Which, when taken at flood, leads on to the happiest fortune, –
> Leads to the marriage-morn and the orange-flowers and the altar,
> And the long lawful line or crowned joys to crowned joys succeeding, –
> Ah, it has ebbed with me! Ye gods, and when it was flowing,
> Pitiful fool that I was, to stand fiddle-faddling in that way! (IV, 33–8)

---

[10] See Clough's *Correspondence* I:278.

Claude looks up at the mountains and imagines that Mary is somewhere up there, calling to him: "Ah, could I hear her call! Could I catch the glimpse of her raiment!" (50). He hurries back to Florence, hoping to get news from Miss Roper and her brother. Meanwhile, Mary is writing Miss Roper from Lucerne, reviewing the missed connections and miscommunications and wondering whether her friend has news about Claude, but she is losing hope: "Well, he is not come; and now, I suppose, he will not come" (63).

Unfortunately, as the reader learns in the concluding Canto V, by the time Claude arrives, the Ropers are no longer in Florence, having gone on to the Lucca Baths, west of the city. Claude pursues rumors about the Trevellyns' whereabouts, traveling to Pisa and then back to Florence, but to no avail. In the anticlimactic ending, Claude struggles to interpret this episode in his life and to reconcile his potential but thwarted union with Mary in terms of his world view. In his opening letters, he still expresses the hope that Eustace will be able to contact certain friends who may be able to furnish information about the Trevellyns, but he gradually resolves to forget her: "Let me, then, bear to forget her, I will not cling to her falsely" (V, 51), and he acknowledges that she was "Worthy a nobler heart than a fool such as I could have given" (69). The intensity of Claude's emotions is seen in his drafts of letters he does not actually intend to send – this is writing as therapy, not communication with his friend. Among his admissions is that by giving up his pursuit of Mary he is "a coward, and know it. / Courage in me could only be factitious, unnatural, useless" (84–5). Rather than try to find a religious or philosophical basis for his decisions, he "will look straight out, see things, not try to evade them: / Fact shall be fact for me" (100–101). As he engages in this self-analysis, he learns that Rome has fallen to the French invaders, while "I, meanwhile, for the loss of single small chit of a girl, sit / Moping and mourning here – for her, and myself much smaller" (116–17). Clough clearly intends to link the emotional impact of this political tragedy – which Claude sincerely cares about but in which he is not personally involved, is incapable of becoming involved in a meaningful way – with the tragedy of his inconclusive affair with Mary. The fall of Rome seemed inevitable and is not surprising, and furthermore Claude is an Englishman, not an Italian, but that does not mean that Claude's non-involvement is fully justified. Claude did in fact pursue Mary with the sincere intention of proposing marriage, but only after his "shilly-shallying" cost him the opportunity to take advantage of the right moment. Why will he not now resume his search? In letter viii, he writes Eustace that he "cannot stay at Florence, not even to wait for a letter" (141), that "I am more a coward than ever, / Chicken-hearted, past thought" (144–5) and, "After all do I know that I really cared so about her? / Do whatever I will, I cannot call up her image" (156–7).

In the final analysis, the complications resulting from the Trevellyns' flight from Rome do not necessarily mean that "Fate" has determined Claude's destiny. After suggesting that "Great is Fate, and is best. I believe in Providence, partly. / What is ordained is right, and all that happens is ordered" (176–7), he reverses himself and writes, "Ah, no, that isn't it. But yet I retain my conclusion: I will go where I am led, and will not dictate to the chances" (177–8). Gone is that "moment" when he responded to impulse, suspended intellectual analysis, put aside doubts and thought beyond himself. The implication here is that Claude is accepting himself – and his life – with their limitations. His final letter comes from Rome, which "will not suit me, Eustace; the priests and soldiers possess it" (186), and before announcing his intention to travel eastward to Egypt, he muses that "Faith, I think, does pass, and Love; but Knowledge abideth. / Let us seek Knowledge" for "Knowledge is painful often; and yet when we know we are happy. / Seek it, and leave mere Faith and Love to come with the chances" (201–2).

It is interesting that the final letter is written not by Claude but rather by Mary, resigned to Claude's resignation, who tells Miss Roper that "You have heard nothing; of course, I know you can have heard nothing" and reckons that Claude:

> Finding the chances prevail against meeting again ... would banish
> Forthwith every thought of the poor little possible hope, which
> I myself could not help, perhaps, thinking only too much of;
> He would resign himself, and go. I see it exactly.
> So I also submit, although in a different manner.
>   ... We go very shortly to England (211–16)

The final epilogue asks, "*Go little book! thy tale, is it not evil and good?*" (V, 218). The hopefulness of the opening prologue, with its references to "a land wherein gods of the old time wandered, / Where every breath even now changes to ether divine" (I, 3–4) has been accommodated to the point of view of the disillusioned idealist. Is the ideal past of the "Eternal City" an illusion?

Clough had every reason to expect mixed reactions to *Amours de Voyage*. His protagonist is a kind of anti-hero, and his "tragic-comedy, or comi-tragedy" is a kind of anti-romance. After Clough had completed a draft of the first version of the poem in October of 1849, he sent a copy to his old Oxford friend John Campbell Shairp, who wrote to Clough about his negative reaction – "everything crumbles to dust beneath a ceaseless self-introspection and criticism" – and strongly advised him not to publish the poem. Clough responded that he did not intend to publish the poem in the near future but thought his friend had not appreciated Claude's "final strength of mind" as his character develops in the narrative (*Correspondence*, 274–8).[11] As already mentioned, when the poem was finally published in *The Atlantic Monthly* nine years later, Emerson himself responded negatively to the unconventional ending: "How can you waste such power on a broken dream? Why lead us up the tower to tumble down?" Clough in his answer to Emerson made it clear that he had always meant to end this poem the way he did, implying that it was right for his protagonist in this life situation (*Correspondence*, 548). He had had nine years

---

[11] Not surprisingly, when Clough sent the poem to Matthew Arnold a few years later, he was not favorably impressed with the poem, either – "what is to be said when a thing does not suit you?" See Arnold's letter to Clough, 21 March 1853, in *The Letters of Matthew Arnold to Arthur Hugh Clough*, 132.

to think about his unusual poem and he remained dedicated to his original intent, despite the criticism that he knew he could expect. Michael Timko sees Clough's "main purpose" as "the exposé of a self-centered prig unable to realize ... the necessity of striking a balance between theory and practice, between independence of and unity with his fellow men" (138). Those who label the ending inconclusive or anti-climactic miss the point that "Clough's purpose has been to show the basic inability of a person like Claude, aspiring somehow to live above the realities of life, to make any kind of compromise with the 'vulgar' world" (139).

Stephanie Markovits interprets Clough's work in terms of what she sees to be a trend in Victorian literature, a "crisis of action": "Writers skirmishing over the relative roles of action and character in literature redefined heroism and influenced the development of the novel as a genre concerned with character and states of consciousness rather than deeds" (446). His poem incorporates "the skeleton of the epic, the immediacy and inwardness of the lyric, and the colloquialism of the novel" (468). For Markovits, Clough's poetry corresponds to an increased emphasis on "internal action" in the Victorian "novel of character, as distinguished from the novel of plot" (477), as Anthony Trollope and George Eliot prepare the way for Henry James, but it is also the case that Clough's psychological realism and anti-romantic stance can be usefully compared to George Meredith's 1862 poem *Modern Love*, and several critics have noticed that the character Claude in his introspective self-analysis anticipates T. S. Eliot's more absurd and ineffectual Prufrock, an ironic observation given comparisons between the indecisive Claude and Shakespeare's Hamlet by nineteenth-century reviewers and by Markovits as well (475–6).[12] In any case, the critical consensus today is that Clough's *Amours de Voyage*, while grounded in its historical setting, is forward-looking in its aesthetics of doubt and uncertainty.

In his own day, Clough may not have anticipated the current evaluation of his three long poems – *The Bothie of Toper-na-Vuolich*, *Amours de Voyage*, and *Dipsychus* – as important contributions to Victorian literature, although he was encouraged by the response to *The Bothie of Toper-na-Fuosich*, the original title,[13] when it was published in 1848. Placing *Amours de Voyage* in the context of the other two poems helps us to understand Clough's continuous efforts to find an appropriate narrative voice and relate his narrative to the universals of human nature. *The Bothie* is odd and eccentric but "happy," a surprise to many who expected something somber from a man who had recently resigned his fellowship

---

[12] Timko refers to the comparison between Claude and Hamlet as "unfortunate" and observes that he is much closer to Eliot's Prufrock (140). See also Biswas (299, 309). In this context, one of the best known lines from Eliot's famous poem acquires a new meaning: "No! I am not Prince Hamlet, nor was meant to be."

[13] Soon after the poem was published, Clough learned from a review in *The Literary Gazette* (18 August 1849) that the original Gaelic title, translated as "the hut of the bearded well," probably refers to a woman's vagina, and, embarrassed, he revised the title for subsequent (posthumous) publication, choosing a similar but innocuous (and meaningless) phrase.

at Oxford. Indeed, the setting nostalgically recalls Oxford reading parties in the Scottish Highlands in which Clough and his friends had participated during the long vacation. The curious hexameter meter, with its heroic associations, was probably inspired by Henry Wadsworth Longfellow's use of it in his recent *Evangeline* (1847), though Clough adapts it for his own purposes. The protagonist is Philip Hewson, an idealistic student with republican sentiments, and the "reading party" plot dissolves into a kind of romantic idyll as he wanders the countryside and searches for the right woman in his life. Philip is attracted to three women in succession: Katie, a farmer's daughter; the elitist Lady Maria; and, finally, Elspie, who lives in the bothie (hut) of the title and is close to the earth like Katie but also possesses understanding and social graces that make her very special. Philip chooses Elspie as his bride, and the two of them leave for a life of farming in New Zealand. (Clough's close friend Tom Arnold, Dr. Arnold's son and Matthew's brother, had emigrated to New Zealand but later returned.) Regardless of Clough's ironic and playful use of classical epic rhetorical devices in addition to the mock-heroic meter, the overall romantic implications of this narrative poem are not undercut, and the joyful tone of this work obviously figures in the responses of contemporary readers which tended to be positive in spite of the poem's odd generic and stylistic attributes. It is striking that Clough's romantic comedy was followed so closely by the "tragic-comedy" *Amours*, in which he once again alludes to ancient heroic traditions in his adaptation of the hexameter but systematically disappoints, through the self-conscious voice of Claude, readers' expectations for heroic and romantic fulfillment.

Ambarvalia, a collection of short poems by Clough and his friend Thomas Burbidge, was published in early 1849, shortly after *The Bothie*, but probably all of Clough's poems grouped here had been written before the concerted effort that produced the long poem. As critics have pointed out over the years, Clough's poems represent his "questioning spirit" (his own phrase) during the Oxford years, and the consensus is that Burbidge, though Clough's friend, is not an important poet. Among the poems by Clough in this collection that are most often discussed, "Natura Naturans" is of particular interest here. The title is a medieval Latin phrase meaning, loosely, "nature doing what nature does," and it is the natural sex drive that the poet has in mind. A man and a woman are riding in the same railway carriage:

> Beside me,--in the car,--she sat
>   She spake not, no, nor looked at me:
> From her to me, from me to her,
>   What passed so subtly, stealthily?
> As rose to rose that by it blows
>   Its interchanged aroma flings;
> Or wake to sound of one sweet note
>   The virtues of disparted strings.
>
> Beside me, nought but this! – but this,
>   That influent as within me dwelt

> Her life, mine too within her breast,
>   Her brain, her every limb she felt;
> We sat; while o'er and in us, more
>   And more, a power unknown prevailed,
> Inhaling, and inhaled, – and still,
>   'Twas one, inhaling or inhaled.
>
> Beside me, nought but this; – and passed;
>   I passed; and know not to this day
> If gold or jet her girlish hair,
>   If black, or brown, or lucid-grey
> Her eye's young glance: the fickle chance
>   That joined us, yet may join again
> But I no face again could greet
>   As hers, whose life was in me then. (1–24)

The speaker goes on to speculate about the biblical origins of human sexuality:

> Such sweet preluding sense of old
>   Led on in Eden's sinless place
> The hours when bodies human first
>   Combined the primal prime embrace
> Such genial heat the blissful seat
>   In man and woman owned unblamed
> When, naked both, its garden paths
>   They walked unconscious, unashamed;
>
> Ere, clouded yet in mistiest dawn,
>   Above the horizon dusk and dun,
> One mountain crest with light had tipped
>   That Orb that is the Spirit's Sun;
> Ere dreamed young flowers in vernal showers
>   Of fruit to rise the flower above
> Or ever yet to young Desire
>   Was told the mystic name of Love. (73–88)

I quote this poem at length because the chance meeting between these two in the railway carriage – which leads to nothing beyond that – so powerfully evokes the idea of *juxtaposition* that is central to *Amours de Voyage*. Clough's preoccupation with the idea that humans mate as a result of chance encounters is significant. The encounter between Philip and Elspie in *Bothie* leads not only to sex, but love, and, beyond that, to an idealized marital union that acknowledges a genuine intellectual as well as emotional partnership, while the potential union of Claude and Mary in *Amours* after frustrating complications remains unrealized. In both cases, however, the central relationship, potential or unrealized, is framed by an unstable or shifting national, social, and familial network. Philip takes his wife from a setting dominated by close-knit familial or clan relationships to a destination halfway around the world connected to their homeland in a broad sense by British imperialism, and Claude contemplates a union with a woman and her English family in an international European setting.

In terms of Clough's composition, *Amours*, at least the first draft of that work, followed hot on the heels of the popular *Bothie*, but, as already discussed, was not published for another nine years, late in the poet's life, and the American serial publication was little noticed in England, while Clough's notable American friend Emerson disliked it. In the meantime, after returning to London from Rome, Clough assumed his primarily administrative duties as principal of University Hall, though apparently he was unhappy there and often found himself in conflict with the school's Unitarian masters who, for example, insisted on daily prayers that made the agnostic Clough uncomfortable. Nevertheless, Clough maintained his creative energy and worked on yet another major long poem, *Dipsychus*. Though the poem remained unfinished and unpublished during his lifetime, it eventually took its place in his *oeuvre* as his third long poem and last important work. Like the *Bothie*, this poem offers interesting comparisons to *Amours*, though it is in many ways a very distinctive, unusual work. The poem is set in Venice, where Clough took a holiday in the autumn of 1850 and consists of 13 scenes in which the idealistic, "double-minded" protagonist debates issues of religion, love, and life with himself, never able to reach satisfactory conclusions, while he is prodded and mocked by a Spirit. This is done somewhat in the manner of Goethe's *Faust*, although in the first half of the poem the Spirit seems to be offering Dipsychus a kind of worldly compromise between idealism and "fallen nature" before becoming more of a conventional Mephistopheles in the second half. Here I want to emphasize the sexual temptations that are central to the protagonist's dilemma. Dipsychus struggles to resist the temptations of prostitutes strolling the streets of Venice: for the tourist Clough such figures may very well have been associated in his mind with English prostitutes, not only in London but in Oxford as well.[14] Dipsychus attempts to focus on the ideal of "sweet domestic bonds" and "matrimonial sanctities" while the Spirit describes a prostitute's chamber, erotic yet ironically containing an image of the Madonna. Of course, it can be generalized that Clough is intrigued by stereotypical Victorian concerns about sexual morality, but it is important to note the centrality of unusual or problematic sexual unions in his work.

In 1851 Clough met his future wife Blanche Smith, as the remarkably prolific period in which he had composed his three major works was coming to an end. The next year he resigned his unhappy position at University Hall, announced his engagement to Blanche, and sailed for America, where he hoped to find a teaching position or found a school of some kind. In Massachusetts, he renewed his friendship with Emerson and made contact with James Russell Lowell, Charles Eliot Norton, and various other intellectuals, educators, and writers but in the end did not find a position that suited him. Acquiescing to the wishes of Blanche, he returned to England in 1853 and assumed duties as an examiner in the Education

---

[14] For conjectures about Clough's associations with prostitutes, see Christiansen, 24.

Office. After settling into this new position with a barely adequate income, he married Blanche in 1854. In 1857 he also began to dedicate himself to humanitarian efforts, volunteering his services to aid his wife's famous cousin, Florence Nightingale, in her work. However, his literary ambitions faded somewhat after his marriage, and he wrote little new work until the last year of his life, though he would complete his revision of Dryden's *Plutarch's Lives*, begun in America, in 1859, and the serial publication of *Amours* in the *Atlantic Monthly* in 1858 brought him the only money he would ever earn from his poetry. In the next few years, his health began to fail, and he undertook a series of foreign travels in an attempt to recover. When he died in Florence in 1861, his wife and children by his side, he had begun a series of poems entitled *Mari Magno*. Inspired by his 1852 trip to America and loosely modeled on Chaucer's *Canterbury Tales*, the stories are narrated by passengers on a transatlantic voyage. Most of them deal with marriage in one way or another but in a relatively conventional manner that avoids the complexities and uncertainties of his earlier work. One of them, "The American Tale," has the subtitle "Juxtaposition," but in it fears of commitment like those associated with Claude in *Amours* are resolved and the chance "juxtaposition" leads to a happy marriage. For obvious reasons, most critics have found this unfinished manuscript of lesser interest than Clough's earlier work, but in a larger context, it demonstrates a Victorian poet's continuing search for a genre and style that will enable him to produce another long poem, and the turn from experimental and provocative forms to a more conventional one in this case may remind us of Alexander Smith's attempt to recover his poetic reputation in his final published work, the historical epic *Edwin of Deira* (1861), with a more conservative, toned-down style, which critics thought was imitative of Tennyson's *Idylls*, as discussed in Chapter 1.

It is appropriate to discuss Clough's poetic career and reputation here because *Amours* is widely judged to be his most important work, though it is not only a product of his troubled aesthetic vision and life experiences but, in spite of its sometimes humorous tone, a poem about failure. Clough's posthumous standing as an English poet was complicated by his reputation for great natural ability but "unfulfilled potential," his fluctuating and controversial religious opinions, and his complex relationships with influential figures (especially the poet Matthew Arnold), but with the publication of his widow Blanche's collected edition of his poems in 1862, followed by fuller collections in following years, Clough's enduring importance was established. In the early twentieth century, he suffered from the backlash against Victorian authors as did his more famous contemporaries (he was satirized as the typical product of the English public school system by Lytton Strachey in *Eminent Victorians* and routinely associated with the portrait of Rugby in Thomas Hughes's 1857 novel *Tom Brown's Schooldays*) and he became fairly obscure, but in the mid-twentieth century there was a remarkable revival of interest.

In 1951 came the first modern edition of the poems by H. F. Lowry, A. L. P Norrington, and F. L. Mulhauser (followed by Mulhauser's edition of

Clough correspondence in 1957), and an outpouring of critical studies in the 1960s and 1970s included Isobel Armstrong's *Arthur Hugh Clough* (1962), W. E. Houghton's *The Poetry of Clough* (1963), Michael Timko's *Innocent Victorian: The Satiric Poetry of Arthur Hugh Clough* (1966), Wendell V. Harris's *Arthur Hugh Clough* (1970), Evelyn Barish Greenberger's *Arthur Hugh Clough: The Growth of a Poet's Mind* (1970), Robindra Kumar Biswas's *Arthur Hugh Clough: Towards a Reconsideration* (1972) and Michael's Thorpe's *Clough: The Critical Heritage* (1982). The most important biography was Katharine Chorley's *Arthur Hugh Clough: The Uncommitted Mind* (1962). A second edition of the poems, edited by Mulhauser, came out in 1974, and in that same year there was an extensively annotated edition of *Amours de Voyage* by Patrick Scott, the edition to which I refer in this study. A consistent theme in Clough studies from this period is an acknowledgment of his intellectual honesty and forthright, emotional engagement with the everyday world while retaining a sense of moral idealism, though alienated from religious belief. Again and again, Clough's champions sought to dispel the "myth" of his "failure." Armstrong perhaps spoke for the majority of literary historians and critics when she described his work as "unorthodox, inventive, sophisticated and self-conscious" and judged that "One cannot claim that he is a major poet; but his unusual and individual achievement makes him among the most exciting and rewarding of Victorian minor poets" (8). Houghton went a bit further: "I will not argue that he was a major poet and not a minor one... I will only claim that for us he belongs with Tennyson, Browning, Arnold, and Hopkins, intrinsically and relevantly" (225). Since Houghton wrote this, it has become especially common to compare Clough's achievement as a poet favorably with that of his friend and rival Matthew Arnold.[15] Though the critical outpouring of this period was not matched in subsequent decades, recent criticism points to a continuing interest in Clough,[16] and, most significantly, Anthony Kenny's biography, *Arthur Hugh Clough: A Poet's Life*, appeared in 2005. Clough is popularly known for a few short poems often anthologized, especially "The Latest Decalogue," which satirizes not the Ten Commandments but the way they are popularly interpreted (e.g., "Do not adultery commit; /Advantage rarely comes of it") and the quite different "Say Not the Struggle Nought Availeth," which has become a classic expression of hope and fortitude, memorably quoted by Winston

---

[15] Kenny writes, for example, that though Blanche Clough did "a disservice to Arthur's memory by fostering the legend that his *oeuvre* was inadequate to his talents," her posthumous editions of his poems in fact demonstrated that Clough "deserved to be ranked among the major poets of the century," and "Most obviously, he compared well as a poet with Matthew Arnold – undoubtedly in the quantity of his verse, and arguably in its quality" (286).

[16] In her 2003 article "Why Clough? Why Now? (Arthur Hugh Clough)," Vanessa Ryan refers to "three explanations for reexamining Clough: first, a revivalist impulse to restore to the canon a poet who has fallen out of interest; second, an interest in the formal aspects of Clough's poetry that may reflect renewed attention to literary form; and third, the prevalence of themes in Clough's poetry that are of central interest to cultural and new historicist criticism" (504).

Churchill in a 1941 radio broadcast in which he asked for U.S. entry into the Second World War. Nevertheless, his poetic reputation rests primarily on the three major long poems, especially *Amours de Voyage*. Biswas, in the introduction to his very substantial study of Clough's life and poetry, makes the point that "it is chiefly in one poem, *Amours de Voyage*, that Clough confidently fulfils his potential as a poet. His success is not wholly confined to this single work, but nowhere else does he achieve this same kind of distinction" (5).

Clough scholars routinely point out that his poetry should be judged on its own merits rather than in the context of his problematic (and sometimes enigmatic) life story, but even Houghton, who asserts that he is concerned with the poetry for its own sake rather than "as a biographical document, or as a record of his thought, or as an index to the age" (xi), is compelled to discuss, for example, biographical information about Clough's 1849 stay in Rome when analyzing *Amours de Voyage*. Beyond Clough's much-discussed relationships with members of the Arnold family, one of the most controversial dealings with his private life is Chorley's psychoanalytical speculation about Clough's exceptionally intense dependence on his mother as a child.[17] It seems reasonable to me that, aside from any esoteric theories, Clough's separation from his mother and father (residing across the sea in America) at the age of nine when he entered Rugby might very well lead to anxieties. More relevant here, however, is commentary about Clough's relationship with his wife, whom he married shortly after completing his first draft of *Amours*.

Among the most frequently quoted letters in Clough's correspondence is one that he wrote to his future wife on January 1, 1852. It was the first after their engagement. Referring to the apple that tempted Eve in the Garden of Eden, Clough interpreted it as representing the belief that love is everything:

> Women will believe so, and try and make men act as if *they* believed so, and straightway, behold, the Fall, and Paradise at an end etc. etc. Love is not everything, Blanche; don't believe it nor make me pretend to believe it. '*Service*' is everything. Let us be fellow-servants. There is no joy nor happiness nor way nor name by which men may be saved but this. (I: 300)

Clough's tone here is reminiscent of Romney Leigh's in the ineffectual early stages of his courtship of Aurora Leigh, and as Kenny points out, Blanche referred to this afterwards as "the terrible letter" (237). Despite such awkward incidents, however, it would be a mistake to identify Clough with Claude in *Amours*. As noted earlier, Clough in fact skillfully employed the conventions of romantic comedy in the *Bothie* just prior to *Amours*. And though some critics have blamed Blanche for Clough's diminished literary output during his married years, others have pointed out that much of his best work had been generated in times of personal troubles

---

[17] Chorley, for example, refers to a "pervasive sense of guilt" associated with Clough's childhood "longing … for complete and exclusive possession of his mother" (352). Later scholars generally do not share her extreme view on this matter, but Chorley's monumental biography is nevertheless an important contribution to Clough scholarship.

and emotional stress. Kenny judges that the marriage was probably a happy one (264). I make this point here because I want to emphasize once more that although *Amours* clearly grew out of Clough's own experiences in Rome, his insights regarding gender and his portrayal of Claude do not necessarily represent some kind of autobiographical revelations. Claude is a witty, mocking doubter, but, as Biswas remarks, Clough was "never entirely at ease in the rebel's role" (305). I would argue that one compelling reason that Clough was not at ease in this role is that although he was painfully aware of his historical situation and especially contemporary intellectual speculations regarding "life philosophy" and religion, he felt grounded as a poet in the predispositions associated with human nature that I have emphasized in this study. The contrast that Biswas finds between "continually recurring images of solidness and fixity and of fluidity and movement" in *Amours* (314) is evidence of Clough's attempt to reconcile the contemporary politics of Rome with the universal themes of human nature.

In my discussion of the poem, I have tried to show in some detail how Claude's philosophical uncertainties and tentative commitments are related to questions of masculinity. At the center of his consciousness, meaning is constantly in process; but, although Clough scholars through the years are quite right to call attention to the emphasis on intellectual ideas – Biswas, for example, refers to "the depth and complexity of the poem and its scrutiny of contemporary history and contemporary values" (321) – human sexuality and gender identity are always implicit not only in the "novelistic," "romantic" plot but in the recurrent images of "cosmic vitality." What we might call his ambivalent manhood is central to his character. Claude understands that a man is expected to take the initiative in courtship, yet he hesitates. He is acutely aware of the tensions between "vile" impulses and moral idealism, between the drive for sex and expectations of social commitment related to the roles of husband and father, between self-preservation and expectations of group solidarity and patriotism, between an intellectual search for truth and allegiance to inherited traditions, between individuality and group identity in a large context. This Roman episode in his life story is revealing because the choices of moving beyond "juxtaposition" to permanent bonds with individuals and groups are placed in a context wherein his identity as an Englishman, with only loose ties to his countrymen and vaguely defined loyalties in a dramatically charged but fluid and unstable political and social milieu, highlights his honestly uncertain values and beliefs. Claude understands that he is being seen as selfish and acknowledges his own cowardice while he takes the opportunity to analyze his relationships with the Roman republic, with his English countrymen, with Mary and her family, in a manner that is invited by the peculiarly stimulating international setting, which is exciting and dangerous and somehow liberating at this historical time while being grounded in ancient cultural and historical traditions. In his own way Clough makes use of his Italian setting as Barrett Browning would do in her way and Browning would do in his. In contrast with the Brownings, however, Clough rejected the tradition of Victorian–Romantic language as he produced what Biswas describes as "a disturbingly unevasive, witty, and complex study of the psychology of the radically self-conscious intellectual" (299).

Claude does not simply abandon his perceived allegiances and responsibilities as an English gentleman – he is "useful" to the Trevellyn family during the French attack of April 30 – but he is contrasted with George Vernon, who has entered into a domestic alliance with the family of his fiance Georgina and who challenges the "intentions" of Claude regarding her sister Mary. In the end, Claude can dedicate himself only to a life of thought. We do not have to identify Claude with Clough in order to see that Claude's "failure" to unite with Mary is intimately connected with Clough's "failure" to conform his narrative poem to the genre of romantic comedy, and it is revealing to contrast Claude's decision to give up in his search for Mary after encountering (real and extended) problems in making contact with the Trevellyns with the heroic journey of a blinded Romney to find Aurora and Marian in Florence.

In general, Clough has gone much further than Tennyson and Barrett Browning and Browning in distancing himself from the poetic language of "Victorian Romanticism." The strong sense of selfhood portrayed in Claude is intellectualized and divorced from Romantic visions while retaining a powerful nostalgic desire for universal meaning apprehended through classical art and cultural artifacts and acknowledging an "instinctive" natural drive not just for sex but for meaningful familial and social bonds and recognition of masculine identity and status. As discussed earlier, literary Darwinists like Carroll argue that "domain-general intelligence has an adaptive function; it facilitates a flexible response to a variable environment" ("Human Nature and Literary Meaning," 87). In these terms, it can be said that Clough's intellectual narrator Claude relies strongly on the cognitive "behavioral system" in analyzing his environment, interpreting history, and telling his personal story. Intellectual openness to experience is central to his character. He is consciously aware of the need for survival and in a humorous but sincere manner acknowledges his "cowardice" in declining to risk his life for either the Roman Republic or fellow English tourists: even English women, even a woman in whom he has taken a special interest. From his male point of view, he acknowledges the "normal" human desire for mating, parenting, the establishment and maintenance of kin relations and social relations. The distinctive setting of Clough's narrative allows the English poet to emphasize Claude's consciousness of mortality in the context of the French invasion and violence in the streets. As argued above, Claude's status as a tourist facilitates his intensive analysis of human bonds. How are sexual attraction and "falling in love" integrated in a meaningful way as "juxtaposition" leads to permanent commitments? What does it mean to integrate into an English family and marry one of its daughters? (It might be interesting to consider, let us say, Jane Austen's novel *Pride and Prejudice* – an extraordinarily popular classic and a particular favorite of Carroll and the literary Darwinists[18]

---

[18] In particular see Carroll's groundbreaking book *Evolution and Literary Theory* (1995) and his essay "Human Nature and Literary Meaning" (2005). In Chapter 1, I refer to a new study by Carroll and others that addresses readers' responses to Austen's novel.

– as an exemplary model in framing the conventions of the nineteenth-century romantic comedy that Clough implicitly questions.) Is parenthood a distinctive role with meaning beyond "avuncular" ties to children? What are the conventions – and the inherent duties – of bonding with one's fellow countrymen and with fellow adherents of a righteous, idealistic political or social cause?

The thoughtful, analytical Claude contemplates these issues in a serious and yet playful mode, and his readers – from Clough's contemporary audience to the (mostly academic) audience of the early twenty-first century – cannot escape the implication that Clough himself is simultaneously satirizing Claude and asking them to think about the implications of Claude's open-ended questions and personal dilemmas. More specifically, there can be little doubt that Clough's own questions and personal dilemmas, especially his long and painful confrontations with the religious controversies at Oxford and deep yet problematic personal ties with family members, friends and mentors, were incorporated into this unique narrative poem. Clough understood the fundamental significance, yet potential volatility, of family ties, the appeal of religious affiliations for which one would be willing to dedicate one's life in tension with the morally sanctioned search for objective truth. The distinctive historical setting of *Amours* allowed him to detach his protagonist in interesting ways from culturally determined identities and relationships that would serve to highlight Claude's *flexible* cognitive resources, the somewhat detached, intellectual consciousness of the *tourist*. Claude, whom Clough surely intended to be somewhat annoying even to fellow English intellectuals, is not presented as an altogether exemplary character, and he does not provide the reader with the kind of powerfully emotional imagery offered up by Barrett Browning's Aurora. Instead, there is a consistent ironical distance implicit in his descriptions. However, he is sincere in his search for truth and apparently honest in his sophisticated yet awkward and oddly constructed personal narrative as related to his friend Eustace. The troubling question implied by *Amours* is whether it may be the case that we are all in a sense "tourists," that the emotionally charged meanings we apply to our most important affiliations in life may be – at a philosophical level – arbitrary, based on "juxtaposition." This helps to account for the troubled, problematic response of the poem's first readers and, at the same time, its continuing appeal. Claude's sincere ambivalence toward his role as an English *man* in revolutionary Italy is conveyed in a verse form that simultaneously suggests a classical tradition of mythic manhood and facilitates a satiric, colloquial, nineteenth-century English expressiveness. Clough's portrayal of the "anti-hero" Claude remains of considerable interest in this sense, and the peculiar genre of the poem contributes to this interest. However, though he "novelized" his narrative poem, Clough's method did not provide a narrative context that would allow the reader to experience vicariously and visualize, for example, the interactions of Claude with Mary and the Trevellyn family, the kind of social context that is essential to the Victorian novel and is to a much greater degree seen in the poems by Tennyson, Barrett Browning, and Browning featured in this study.

Chapter 5
Browning's Chivalrous Christianity

Robert Browning's poetic career is distinctive in three ways that are especially significant in the context of this study. First, among the major Victorian poets, Browning was most prolific in his composition of long poems, and *The Ring and the Book*, often described as a kind of epic, is one of the most discussed and highly regarded long poems in the English language. Second, Browning's relationship with his wife is important to the poem in terms of thematic issues that go beyond biographical "background" information. Third, apart from Browning's relationship with his wife, an emphasis on gender and – of special interest here – complex themes related to masculinity, are central to his work as a whole. Browning's *Men and Women*, a collection that contains poems that eventually took their place among his most famous works – including "Fra Lippo Lippi" and "Andrea del Sarto" – was published in 1855, two years before *Aurora Leigh*, but this bit of publication history in itself can be misleading about his poetic career compared with that of his wife during the years of their life together, 1846–61. Despite what was apparently a mutually happy relationship and Barrett Browning's improved health and impressive literary productivity after their move to Italy, Browning himself wrote relatively little in the early years of their marriage and was largely disappointed by the critical reception of the work he was able to publish: even reviews of *Men and Women* were generally negative and discouraging.

Although Browning had traveled abroad – including a journey to Russia in 1834 – most of his time had been spent living in his parents' home at Camberwell, near London, where, in a comfortable and familiar environment, he had been able to concentrate on his literary efforts. His anonymous long poem *Pauline* (1833), of interest to later critics primarily because it dramatizes the strong influence of Shelley in Browning's work, was not successful. *Paracelsus* (1835) and *Sordello* (1840) found small audiences, and the latter work in particular helped to establish Browning's reputation for obscurity that would remain even after he achieved considerable success in his late career. His six plays during the period 1837–46 were all failures, but his powerful ambition to write for the theater was gradually channeled into the creative development of dramatic poetry in new ways. His experimentation led to *Pippa Passes* (1841), the first of a series of poetic pamphlets called *Bells and Pomegranates*, and the collection *Dramatic Lyrics* (1842), also part of the series and already representative of his mature work with the dramatic monologue. The 1842 volume contains poems that are still among the best known and most widely anthologized of his works, including "Porphyria's Lover," "Soliloquy of the Spanish Cloister," and "My Last Duchess." It is appropriate that any study of masculinities in Browning's poetry take special notice of "My Last Duchess," which became one of Browning's most frequently anthologized poems and one the most popular Victorian poems of all. Browning probably modeled this classic portrait of an aristocratic male domestic tyrant on Alfonso II, fifth and last duke of Ferrara (1533–97), whose young bride Lucrezia died under mysterious circumstances in 1561.[1] In the original edition, the poem was entitled "Italy" and paired with a companion poem, "France," which was later renamed "Count Gismond." Apparently Browning had meant to contrast what he took to be representative treatment of women in the two national traditions: as a mere possession or beautiful object in the first, as a vessel of innocence whose honor must be defended at all costs in the second. In obvious ways the Duke's monologue anticipates Browning's portrayal of other tyrannous husbands, including Guido Franceschini, the Florentine nobleman of depleted fortune who functions as the villain of *The Ring and the Book*.

Browning continued the *Bells and Pomegranates* series until 1846, the year of his marriage. His first collected edition of poems was published in 1849, and that year was also notable because of the birth of Browning's son Pen and the death of his beloved mother in England. The religious poem *Christmas-Eve and Easter Day*, published the following year, did little to enhance his critical reputation, and, in his poetic career, Browning made little progress during the decade 1846–55.

In addition to the momentous changes in Browning's personal life, the unsettled but exciting political environment of his new home and the process of making new social contacts within the community of the English living there absorbed his attention and probably contributed to his difficulties in settling into work habits that would enable him to produce the popular and critically recognized poetry he longed for. Over the years, a great deal has been written about the romance between Robert and Elizabeth, but abundant evidence from personal correspondence and other sources supports the generalization that Browning considered his wife to be a great poet, more talented than himself, and he strongly supported her in her work. After all, according to his first letter to her, he had fallen in love with her on the basis of her poetry, prior to their first meeting, but that did not mean he was not ready to love a real woman. Beyond a sincere appreciation of her poetic art, Robert's idealization of Elizabeth as a woman is reflected in letters in which he expresses his desire, even the need, to "look up to" and "obey" the woman he loves, and Elizabeth's responses occasionally show her uneasiness with his feeling that, from his point of view, there could be no love except "*from beneath*."[2] During the time of their marriage, her poetic reputation continued to grow and was always much greater than his, but Robert's assumptions about gender relations obviously figure heavily into his comments about his sense of selfhood and feelings of self-worth in relation to Elizabeth.

When Elizabeth died in June of 1861, Robert expressed his grief profusely and from that time onward made observances of her memory and his undying love for her an important part of his life that was noted by friends and associates.

---

[1] Alfonso II was the last of the main branch of the Este family. Browning had studied the history of this family as preparation for his long poem *Sordello* (1840). Lucrezia was the 14-year-old daughter of Cosmino I de'Medici, Duke of Florence.

[2] See the exchange of letters between Elizabeth (10 August 1846) and Robert (13 August 1846) in Kintner's edition (953, 960).

He incorporated this theme of veneration for his deceased wife into what would become his most important poetic project of all: in 1864, as he began to work on the massive text that would become *The Ring and the Book*, he not only dedicated it to Elizabeth but made clear her associations with the central image of the ring and identified her with the muse figure he calls "lyric Love, half-angel and half-bird," who inspires him to write. Furthermore, as already suggested in Chapter 1, when he developed the idealized female character Pompilia, he associated her with an idealized image of his deceased wife. This strong "presence" of Elizabeth in the poem is ironic in light of her generally negative attitude toward this poetic project.[3]

In a larger sense, it is ironic that while Robert was grieving for the lost love of his life, his reputation as a poet was growing rapidly, and soon after Elizabeth's death he made significant changes in his personal life. Telling his friends that his heart would remain buried in his wife's tomb, soon after the funeral Browning, with his son Pen, moved back to England, and he would never return to Florence. He reasserted his English identity and made it clear that his son, now 12 years old, would be raised as an Englishman. The turning point in his poetic reputation came with the positive reception of a three-volume collection of poems (many of them revised) and a selected edition, both published in 1863. Then in the spring of 1864, the new collection *Dramatis Personae* was a success. It included several poems that have continued to be popular with readers and scholars through the years, including "Abt Vogler," "A Death in the Desert," and "Caliban upon Setebos." Several of the poems commemorated his wife, though in "Mr. Sludge, the 'Medium,'" the speaker discloses not only his own dishonesty and hypocrisy but that of the patrons who supported his career, and the poem seems to be aimed at generally discrediting spiritualism, to which Barrett Browning had been strongly attracted. Also, in the first poems published after her death, Browning even intensified his longstanding focus on dysfunctional marriages. For example, in "James Lee's Wife," the wife speaks in a series of nine lyrics, recording her husband's emotional withdrawal from her, and "The Worst of It" describes a broken marriage from the man's point of view. Biographers have speculated about the possibility that Browning was giving vent to some conflicted feelings toward his own marriage, based on his romantic idealization of his wife in contrast with frustrations arising from the apparent stifling of his own artistic career while hers was flourishing.[4] At any rate, after his successes of 1863 and 1864, Browning's reputation would grow steadily for the rest of his life.

In fact, it continued to grow after his death in 1889, though he appealed to quite varied audiences. In his influential 1903 book on Browning, G. K. Chesterton praised him and his works. Henry James was interested in the way Browning developed his characters and, especially, the relationships between men and women, and he saw a potential novel in *The Ring and the Book*. To some extent, Browning was grouped with other Victorian authors in the reaction against the period by modernist writers

---

[3] See Irvine and Honan, 409.
[4] See Ryals, 146.

and critics, but Ezra Pound acknowledged Browning's influence and praised his development of dramatic poetry. At the same time, popular reading audiences, especially in the United States, were attracted to what they perceived to be Browning's vitality, appreciation for individualism, and strongly held belief in God and Christian religious values though vaguely defined. Individual "Browningites" exerted influence as cultural leaders and newly organized Browning societies flourished, again particularly in America.[5] At both the academic and popular levels, Browning's biography was routinely studied along with his work, and his courtship and marriage to Barrett Browning were emphasized.

Critical approaches to Browning were varied, however, and the period between the world wars was in general less kind to Browning, as newer versions of the old criticism of his crude and grotesque style and characteristically morbid subject matter were developed. In the 1930s, for example, the influential F. R. Leavis praised Eliot, Pound, and Hopkins, who had broken away from Victorians in general, and he was critical of Browning in particular for his supposed lack of intelligence and for his unsophisticated aesthetic style.[6] He traced imperfections in Pound back to the unfortunate influence of Browning. As for views emphasizing Browning's supposedly heroic character, Richard Altick is an example of a literary scholar who questioned the popular image of "Browning the optimist" and saw instead a man and artist who concealed his disappointment and disillusionment and who was besides a "philosophical illiterate."[7] In a 1952 biography that shocked many of the poet's admirers, Betty Miller, influenced by Freudian psychoanalysis, presented a neurotic Browning who was somewhat overdependent on his mother and less than an ideal partner for Barrett Browning, who was already morbid in her own way. At the same time, influential Browning defenders like William C. DeVane continued to present a much more positive version of the man and his work.

Robert Langbaum's 1957 book *The Poetry of Experience: The Dramatic Monologue in Modern Literary Tradition* was particularly influential in restoring Browning's eminence as a major poet. Langbaum emphasized the central importance of Browning's generic innovations and his role in developing the Romantic literary traditions he inherited in new, modern ways that assumed moral ambivalence in the reader. The key text emphasized by Langbaum was *The Ring and the Book*, but many of Browning's monologues were increasingly interesting to the New Critics. More traditional admirers of Browning continued to insist on his advocacy of Christian values and objected to the implicit relativization of morality in mainstream literary criticism, but Park Honan's *Browning's Characters: A Study in Poetic Technique* (1961) reinforced the work of Langbaum and various other critics, strengthening the "Browning industry" in the 1960s, when there was still a significant overlap between academic and popular audiences. However, Browning's poetry, especially *The Ring and the Book*, continued to be of academic interest, even as academic audiences became increasingly specialized and

---

[5] See O'Neill's chapter on "The Browningites," 1–28.
[6] O'Neill, 68.
[7] See Richard Altick's "The Private Life of Robert Browning" (1951).

increasingly isolated from more general or popular audiences. In the late 1960s and 1970s, Isobel Armstrong and other influential critics led the way in encouraging ever more diverse approaches to studying Browning's work.

In the 1970s and 1980s, literary theorists from a variety of perspectives – structuralists, deconstructionists, phenomenologists, reader-response critics, new historicists, advocates of cultural studies, and others – continued the trend of applying new methodologies to Browning. (Feminists did not have a great deal to say about Browning, but in my discussion below I refer to a few studies that are of interest in considering Browning and gender.) Herbert F. Tucker's *Browning's Beginnings: The Art of Disclosure* (1980) explicitly seeks to place Browning in terms of modern theory and emphasizes the poet's constant process of striving, becoming, and avoiding closure. Tucker's book and others, such as E. Warwick Slinn's *Browning and the Fictions of Identity* (1982), helped to open up Browning to postmodernists in the following decades. In his contribution to the special edition of *Victorian Poetry* dedicated to Browning in 1989, W. David Shaw generalized that "[p]roblems associated with contemporary deconstruction and hermeneutics were familiar to Browning and already understood by him in his Roman murder mystery, *The Ring and the Book*" (79). Other critics sought to combine postmodernist approaches with a historical understanding of Browning in his historical context. Armstrong, for example, in her 1993 book on Victorian poetry, examined Browning's poetry as cultural critique.

At the same time, a more traditional combination of scholarship and "human interest" invested in the Brownings – both as individual artists and in their life together – was still evident, notably that associated with the Browning Library, established in 1918 on the campus of Baylor University, and in 1951 renamed the Armstrong Browning Library in honor of its founder, A. J. Armstrong. Overall, there remained a remarkable level of interest in these poets' personal lives.

When he returned to England following his wife's death in 1861, Browning apparently enjoyed his newfound celebrity, leading an active social life. He had few monetary concerns because of the bequest made by John Kenyon, as noted in Chapter 3, and his increasing fame gave him the confidence he needed to sustain his efforts during the years 1864–68 on what would become his most substantial and important poetic project. The circumstances of Browning's discovery of what came to be known as the Old Yellow Book are well known. An old book with vellum covers that he had purchased at an outdoor market in Florence's Piazzo San Marino in June 1860 contained printed and manuscript materials related to a trial that had taken place in Rome in 1698. The basic facts were uncontested: Guido Franscheschini, an impoverished nobleman and minor church official from Arezzo, along with hired accessories, had murdered his (teenage) wife and her parents. The motive for the crime was particularly interesting to Browning. Francheschini claimed that his young wife was having an illicit affair with a local priest, and it was "disputed whether and when a Husband may kill his Adulterous Wife without incurring the ordinary penalty."[8]

---

[8] See John Marshall Gest's edition of *The Old Yellow Book*, 26. All citations given below are to this edition.

Browning's intuition that this remarkable old book would provide ideal source material for a literary narrative was not supported by his wife, as noted earlier, and this is probably the main reason for his delay in taking advantage of his discovery. He offered the idea to several writers, including the historian W. C. Cartwright, the poet Tennyson, and the novelist Anthony Trollope, but none of them accepted. Browning himself eventually published his long poem based on his Italian discovery in four monthly installments from November 1868 through February 1869. In constructing his *Idylls of the King* over the years, Tennyson took advantage of his increasing stature as a leading poet of Victorian England in offering a mythic narrative that did not directly connect with contemporary history, yet suggested origins of national identity, a symbolic hero, and universal principles of human nature with which his readers could identify. Barrett Browning in *Aurora Leigh* and Clough in *Amours de Voyage* each developed a contemporary protagonist who, although fictional, could be identified with the author in obvious ways. Browning inserted his own poetic voice into *The Ring and the Book*, providing not only a narrative "frame" for the historical documents he asserted were real but a narrative point of view designed to compete with – and dominate – the distinctive points of view brought by other narratives supposedly based on historical characters.

> The "real author" is speaking directly to his contemporary readers, grounding his claims on apparently verifiable data, and making references to his personal life and commitments beyond that of his "reading" of the documents. Furthermore, as Tucker points out, the implicit relationship between the time of the events depicted in the poem and that of the telling

in effect puts the Victorian era on trial. Gender politics, class conflict, the girding of church against state and of orthodoxy against new philosophy – all these were astir in a 1690s Italy that resembles 1860s Britain because, within the pan-European perspective Browning meant to sustain, the epochs were knotted into the same fabric of world history (*Epic*, 440).

The entire poem consists of 12 books, each book consisting of an extended monologue by an individual speaker. In the individual monologues, the story of the tragic marriage and murders is told repeatedly, from different points of view. On the one hand, Browning continues to develop the genre of the dramatic monologue in distinctive and innovative ways; on the other hand, the massive, 12-book structure of his work suggests that of the classic epic.

In book 1, the primary narrator is identified with the poet himself:

> Do you see this square old book yellow Book, I toss
> I' the air and catch again, and twirl about
> By the crumpled vellum covers,–pure crude fat
> Secreted from man's life when hearts beat hard,
> And brains, high-blooded, ticked two centuries since?
> Examine it yourselves! (32–7)

Then, toward the end of the first book, he addresses the "British Public, ye who like me not, / God love you," an obvious reference to his past frustrations in attempting to reach his British audience, but at this point, in the context of his growing fame, it is not a bitter challenge but an ironic and subtle appeal to the readers who have begun to appreciate his work. At the beginning of the book, the ring had been introduced as a metaphor for artistic creation. The artist's imagination and "pure crude fact" form an alloy from which the literary artifact is forged. It is also an allusion to his wife and her poetry: a memorial tablet for Barrett Browning had been placed at Casa Guidi, inscribed with a verse by the Italian poet Nicolò Tommasei in which he refers to her poetry as a golden ring linking Italy and England. At the end of the first book, allusions to the ring and to his muse again recall his deceased wife and solidify readers' identification of the narrator with the poet himself.

In the course of this book, which is entitled "The Ring and the Book," the poet as first-person narrator lays out the details of the case. In 1693 Count Guido married the young Pompilia, daughter of the elderly Roman couple Pietro and Violante Comparini, in the mistaken belief that she was wealthy. Initially, the parents moved into Guido's home in Arezzo with their daughter and son-in-law, but everyone was unhappy and the parents soon returned to Rome. Then Violante – who, rather than her husband Pietro, took the initiative in all matters related to this story – revealed that Pompilia was illegitimate, the child of a prostitute, and on that basis the Comparinis sued Guido for the return of her dowry. After living less than four years in her miserable union with Guido, Pompilia fled from him in the company of Caponsacchi, a young priest. When they stopped to rest at an inn in Castelnuovo, on the outskirts of Rome, the two were overtaken and arrested: Pompilila was taken to a nunnery and Caponsacchi was tried and sentenced to three years of internal exile while lawsuits were being decided. Soon it was discovered that Pompilia was pregnant, and she was released into the care of Pietro and Violante, but on January 2, 1698, about two weeks after Pompilia had given birth to her son Gaetano on December 18, Guido and four accomplices armed with knives entered their villa, murdered the old couple, and fatally wounded Pompilia. After his arrest, Guido claimed that his wife and Caponsacchi had been carrying on an adulterous affair, and thus the killings should be ruled justifiable homicide. He was nevertheless convicted, but then he pleaded exemption on the grounds of a minor office he held in the Church. Guido's sentence of capital punishment was thus appealed to the Pope, but after reviewing his case, the Pope refused to help Guido and called for his execution, which took place shortly thereafter. The poet–narrator describes the creative process by which he first read and interpreted the "raw facts" of the documents bound together in the book, quickly forming his own interpretation of the essential story involving the "monster" Guido, his saintlike, victimized wife, the chivalric priest Caponsacchi, and the others. He goes on to acknowledge the various points of view expressed in the documents and outlines the structure of the poem as a whole.

In book two, "Half-Rome," the speaker is an unhappily married man who sympathizes with Guido. He speaks on the day after the murder of the Comparinis while Pompilia is dying of her wounds. He sees Guido as a victim of the old couple and one who is justified in delivering a "blow of honor" to a faithless wife.

It is gradually revealed that the speaker is quite biased, one who suspects his own wife of being unfaithful, and flourishing a knife, he considers using it to punish "a certain what's-his-name and jackanapes" whom he suspects of being interested in his wife.

Book 3, "The Other Half Rome," is narrated, two days after the murderous attack, by a bachelor sympathetic to the plight of the beautiful, dying Pompilia. He sees the Comparinis as innocent victims as well, and he condemns Guido as an abusive husband and murderer who should pay for his crimes. This narrator seems reasonable and unbiased in his opinions until near the end of his monologue, when he mentions a law case in which both he and Guido were involved. Apparently, Guido had objected to a certain clause in a legal contract and accused the man identified as "The Other Half Rome" of being a liar. It appears that his speaker, like "Half-Rome," is biased in his interpretation of the story.

The speaker in book 4, "Tertium Quid" ("third quantity") offers a third point of view that is apparently unbiased toward either side, but cold and cynical. He reveals himself to be an ambitious man seeking patronage, as he addresses someone identified as "Highness" and another as "Excellency." In contrast to the direct, vernacular style of the second and third books, this speaker is formal and dignified. Although he does not see a clear right and wrong in the story, he is somewhat deferential to Guido's rank and condescending to Pompilia because of her illegitimate birth.

Book 5, "Count Guido Franceschini," is the first of Guido's two monologues. A few days after the attack, and just after release from the torture rack, he is pleading his case before the judges. After implicitly appealing for sympathy in references to painful injuries as a result of being tortured, he eloquently confesses but justifies his actions by making the Comparinis appear wholly treacherous and Pompilia faithless. He refers to alleged love letters between Pompilia and Caponsacchi as evidence of their illicit affair.

In contrast to the previous book, book 6, "Giuseppe Caponsacchi," offers a speaker who is not deferential toward the judges, whom he also addresses. The priest is angry, and his speech is distorted by emotion as he refers to his previous appearance before them after his attempt to rescue Pompilia from her hateful husband. At that time, if they had acknowledged the truth of Pompilia's essential goodness and Guido's essential evil, they could have acted to prevent him from carrying out such a crime. He tells of his own search for a meaningful life in the priesthood, something he never found until he committed himself to help save Pompilia, whom he identifies with the Madonna. He insists that his love for her has been chaste, and he imagines Guido meeting Judas Iscariot in hell.

In book 7, Pompilia speaks for herself from her deathbed and in her own way reinforces much of what Caponsacchi had to say, asserting her own purity and innocence. Describing her relationship with the priest as a spiritual one, she dismisses the supposed love letters as forgeries and remarks that she is unable to read and write. Like the priest, she appeals to God's truth, which should be apparent to everyone; unlike the priest, she is without anger, and she calmly forgives Guido. Her confident, straightforward spirituality is expressed in her dying words, asserting that God shows "His light / For us in the dark to rise by. And I rise" (VII, 1828).

The following eighth book, "Dominus Hyacinthus de Archangelis," makes an abrupt turn, as Guido's attorney prepares his notes for his presentation to the court, arguing that his client's actions are justified by the laws of God, man, and nature. Browning's satire of the glib attorney is obvious: Archangelis is more concerned with his own home and family than the case being argued – in particular, he is distracted by thoughts about his young son. Though he apparently identifies to some extent with Guido's professed defense of his familial honor, it is clear that he is primarily interested in the mechanics of law rather than "truth." However, he is proud of his knowledge of the law and his ability to construct legal arguments.

Browning continues his satire of lawyers in book 9, "Juris Doctor Johannes-Baptista Bottinus," as Pompilia's legal representative is preparing his arguments for the judges. Unlike Archangelis, he is a bachelor and seems totally absorbed in the legal profession. In constructing what he intends to be a tour de force, he begins by reviewing the accusations of adultery against his client as if they were justified and then reverses his arguments and refutes them, displaying his learning and oratorical skill and incorporating insults aimed at Archangelis. Though he represents the spiritual Pompilia, Bottini is no more interested in "God's truth" than Archangelis is.

The Pope comes next, in book 10, alone, pondering to himself the guilty verdict that has been reached by the court and Guido's appeal to him for clemency on the basis of his holding a minor church office. He is temporal ruler of the Papal States where the murders took place, but the Pope finds the legal arguments themselves confusing and unconvincing. Like Caponsacchi and Pompilia, he looks for a transcendent truth beyond the legal intricacies, and he decides that although Caponsacchi's behavior has been rather extraordinary for a priest, clearly it was justified because from the Pope's point of view Guido is a "blotch of black" (X, 865) and Pompilia is "Perfect in whiteness" (X, 1002). Although he begins his monologue by acknowledging the fallibility of his predecessors in making decisions, his sympathetic identification with key players in this life drama, the pure Pompilia and her "warrior-priest," leads him to God's truth, and his decision is clear and firm: he is an elderly man and "how should I dare die, this man let live?" (X, 2127).

In book 11, the longest one, Guido, in prison and awaiting execution, is visited by two high ecclesiastics who have come to offer him confession. Remarkably articulate in this extremity, Guido boasts of his villainy but demands mercy from the Church and clemency from the law. He attacks the basis of his accusers' Christian faith, and he threatens to return after death as a monster of vengeance, attacking Pompilia and Caponsacchi. At the end, he breaks down and desperately begs for forgiveness:

> I am yours,
> I am the Granduke's – no, I am the Pope's!
> Abate, – Cardinal, – Christ, – Maria, – God ...
> Pompilia, will you let them murder me? (XI, 2417–19)

Book 12 is entitled "The Book and the Ring," suggesting a circular or ringlike structure of the work as a whole, and the poet–narrator returns to deliver his concluding monologue and remind the reader that a ring has no beginning or end, though "Here were the end, had anything an end" (XII, 1). He cites letters that describe Guido's execution as well as the restoration of Pompilia's reputation through the influence of the Pope, and he gives details about the contemporary political situation in Rome. He also philosophizes on the elusiveness of truth and the limitations of human knowledge but makes claims for art, which is not life and can tell a truth "obliquely" by addressing mankind rather than individuals and "mean, beyond the facts." He closes by again addressing "Lyric Love" and alluding to the memorial tablet for Barrett Browning, which had been placed at Casa Guidi. Along the way, he addresses his audience in a manner that recalls the introductory first book, phrasing his words in a subtly more optimistic tone –"British Public, who may like me yet" (XII, 831), signaling his confidence in his text.

Since Browning's initial success with *The Ring and the Book* – the *Athenaeum* referring to it as "the *opus magnum* of our generation," "the supremest poetical achievement of our time," and "the most precious and profound spiritual treasure that England has produced since the days of Shakespeare"[9] – the critical consensus has been that it is his masterpiece, as implied by the title of William Irvine and Park Honan's influential 1972 biography, *The Book, the Ring, and the Poet*. In a sense, "gender issues" have always been central to Browning studies in general and discussions of *The Ring and the Book* in particular. The massive scholarship that makes connections between "Browning the man" and "Browning the artist" necessarily emphasizes the relationship between Robert and Elizabeth that begins, prior to their meeting, with Robert's letter announcing that he has fallen in love with her poetry and with the poet as well. As described above, his dedication of *The Ring and the Book* to Elizabeth and the highly foregrounded references to her in books 1 and 12, along with a more subtle parallel between the saintly character Pompilia and his consistently idealized Elizabeth, ensured that the related topics of their romance, marriage, and mutual aesthetic and intellectual influences would be central to discussions of the poem.

The broader topic of "Browning and women" – in the long poem and in his work in general – was also from the beginning of obvious interest to Browning's critics: even those who did not emphasize his personal life and his "uxuriousness" often noticed Browning's tendency to idealize women in his poetry. At the same time, however, they could not ignore the "dark side" of his representations of women as victims. One of the earliest of his dramatic monologues that went on to become part of the canon of Victorian poetry is "Porphyria's Lover," in which the young woman named in the title is dead, having been strangled by her lover, who has killed her, he says, to preserve forever the moment when she was "mine, mine fair, / Perfectly pure and good" (36–7). "My Last Duchess" also features a female victim, of course. In other early poems, women are murdered by husbands

---

[9] Quoted in Ryals, 171.

or lovers or, as in "The Flight of the Duchess," submissive women in repressive relationships are liberated from their tyrannical husbands and enabled to live a new life – in this case, with a band of Gypsies – and the heroine's fulfillment is found not with a chevalier but with an old Gypsy woman in a very feminized setting. Such representations of women were cited by critics who probed into complexities and "problems" in Browning's mind and character and the issue of the "two Brownings," that is, the religious idealist and optimistic champion of heterosexual love, on the one hand, and a disappointed idealist and insecure pessimist on the other. Some critics made specific references to what they saw as Browning's troubled masculinity, referring to his apparent preferences for the company of women,[10] and Miller's psychoanalytic portrait of a neurotic Browning overly dependent on his mother has already been cited. Later, in a 1986 essay on "Browning and Women," Ashby Crowder concluded that "in [Browning's] own life, women were more important to him than men." (134). His generalization is based on a close reading of Browning's correspondence.

Also in the 1980s, but concentrating on Browning's art rather than his biography, U. C. Knoepflmacher pointed out that Browning's conflicted but important "lifelong urge to represent the imaginative possession of a female epipsyche" was related to his struggles with narrative and lyric genres and his development of the dramatic monologue. He shows how the poet's concept of the "female other" in early poems like "Porphyria's Lover" and "My Last Duchess" is central to this artistic development and notes that "[u]nless rescued by the reader, Porphyria and the Duchess remain the perennial captives of masculine speech" (104).

Perhaps the extraordinary interest in the complexities of Browning's "treatment of women" by "traditional" scholars discouraged feminist approaches. William E. Buckler, in his book *Poetry and Truth in Robert Browning's* "The Ring and the Book" (1985), attempted to "redirect" criticism of Browning's classic toward a greater aesthetic appreciation of its greatness as poetry rather than as representing the author's moral, religious, historical, or philosophical positions. Influenced by Henry James's "The Novel in *The Ring and the Book*," Buckler concentrates on Browning's "fusion of fact and imagination that is at once explosive, comprehensible, and nonformulaic" (23) rather than what he took to be the formulaic interpretations of influential critics such as Langbaum, Altick, and Loucks. However, like many earlier critics, Buckler continues to emphasize the centrality of male–female relationships in the poem, though his approach obviously has been influenced by the feminist movement. His commentary is saturated with references to the "theme of a virulent male sexism" (18), and he continually returns to what he takes to be the central issue of men's domination of women in his analysis of the various speakers and their monologues. In my discussion below I refer to some of his assumptions about Browning's "construction of gender," though he does not use that term and generally does not employ the terminology of gender studies.

---

[10] See Altick's essay as an example.

The first substantial monograph offering a professed feminist reading of Browning's poem came in 1988: Ann P. Brady's *"Pompilia": A Feminist Reading of Robert Browning's "The Ring and the Book."* Picking up on Browning's extreme idealization of the character Pompilia, Brady attempted to recruit Browning for feminism, emphasizing his celebration of women and his condemnation of their male oppressors and, by implication, the institutions of patriarchy, including traditional Christianity. Other feminist critics did little to follow up on Brady's work, however, and afterwards the few who expressed an interest in Browning took different approaches. For example, Nina Auerbach, in an article to which I refer in Chapter 1, takes an unsympathetic look at what she believes to be Browning's manipulation of his wife's posthumous reputation by associating her with the character Pompilia in a way that she would not have approved.[11]

Beyond associations with Browning's private life and his relationship with his wife, however, gender issues are of central importance in *The Ring and the Book*. Most obviously, the masculinity of the principal male characters, Guido and Caponsacchi, is at issue in the plot. Interestingly enough, Count Guido Franceschini is routinely referred to as "Guido" by literary critics, whereas the priest Giuseppe Caponsacchi is conventionally referred to by his surname. This is not surprising because Browning gives the count's full name as the title of his first monologue, but only his first name as the second, in which "Guido" (ironically stripped of the markers of social rank and privilege that he prizes) fully reveals his cynical, sadistic, pathetic character. Caponsacchi, a "warrior–priest" (as the Pope describes him) who frames his decision to rescue Pompilia from her abusive husband in terms of his holy mission to serve God, is a remarkable character. Browning no doubt considered his portrayal of a courageous and sexually attractive yet fundamentally moral and religious man as an especially challenging task.

In terms of gender roles, Guido is a complex character as well, however, and not only because of the elaborate rhetorical devices he employs in his attempt to convince his audience of judges to spare his life in his first monologue or in the contrasting style of gushing venom in the second. These much-discussed subtleties in Browning's portrayal of Guido of course reinforce an unsympathetic response in the reader, but the issue of Guido's masculinity in this context has not yet been fully explored. Obviously, even before the murders, his lack of affection for Pompilia and his apparent physical and psychological domination and abuse of her are signs of his deficiencies as a husband and, more generally, as a human being, but I would like to consider his "negative masculinity" in more depth. In particular, I am interested in his use of violence and his attitude toward violence. Although he

---

[11] See Chapter 1, n. 18. According to Auerbach, Browning in *The Ring and the Book* "resurrects his sainted wife in order to butcher her in the person of Pompilia" (168). Other versions of Browning's allusions to his wife in *The Ring and the Book* and the similarities between his poem and *Aurora Leigh* include James McNally, "Touches of Aurora Leigh in The Ring and the Book"; George M. Ridenour, "Robert Browning and Aurora Leigh"; and Flavia Alaya, "The Ring, the Rescue, & the Risorgimento Reunifying the Brownings' Italy."

is utterly ruthless in his willingness to destroy others, he lacks physical courage, and his cowardice undercuts the possibility that his manhood should be respected – and this issue is related to moral judgments but in complex and ambiguous ways. It must be kept in mind that although the character "Guido," like the other principal characters, is loosely based on the historical figure documented in the Old Yellow Book (subsequently cited as OYB), he, like the others, is in the end a creation of Browning's imagination and, as will be noted below, Browning not only creates dramatic scenes for which he has no source but occasionally ignores or contradicts factual information given in the historical documents.

After the first part of book 1, the main incidents of the plot are never in question – one of the most distinctive characteristics of this "narrative" poem – but varying interpretations of these incidents by the individual speakers depends on the way certain details are perceived. In analyzing Guido's masculinity as represented by the speakers, the description of Guido's behavior when he overtakes Pompilia and Caponsacchi at the inn at Castelnuovo is central. Browning changed the date of Pompilia and Capnsacchi's flight from April 28 to April 23 (St. George's Day), 1697. The pair spent the night at the inn (though Pompilia never deviated from her testimony that they had arrived in the morning, shortly before Guido caught up with them and had them arrested). Subsequently, Guido prosecuted them both for flight and adultery, filing law suits first in Rome, then in Arezzo as well. The Roman court decided that Caponsacchi should be exiled to Civita Vecchia for three years and that Pompilia be temporarily restrained at the Monastery of the Convertites. (Browning changed this detail as well – according to the OYB, it was the convent Le Scalette.) The Arezzo sentence was more severe but unenforceable because it was decided that the two were under Roman jurisdiction. Pompilia stayed for three weeks among the nuns and then, because of her pregnancy, she was allowed to move back home to live with her parents. Her son Gaetano was born there on December 18, 1697, and on the following January 2, Guido and his accomplices made their attack.

Throughout the monologues, Guido and his supporters claim that "justice was not served" by the lenient provisions of the Roman court – although they did establish the "guilt" of Pompilia and Caponsacchi, that is, their adultery – and that Guido was therefore justified in taking the law into his own hands. The incident at Castelnuovo is obviously crucial to the plot in preparing the way for the climactic murder scene, and two scenes are repeatedly described, with variations, by the speakers: the initial confrontation, "man to man," between Guido and Caponsacchi in the court of the inn prior to the arrival of the arresting officers, and inside the room, just before she is restrained and arrested, Pompilia's impulsive attempt to draw Guido's own sword in order to attack him. In books 2–11, Browning develops and contextualizes the question of Guido's masculinity in terms of his "cowardice" and related issues.

Half-Rome, sympathetic to Guido, emphasizes the inappropriate (and ridiculous) dress and demeanor of the priest as chevalier: "There strutted Paris in correct costume, / Cloak, cap and feather, no appointment missed, / Even to a wicked-looking sword at his side" (II, 998–1000). But Guido's failure to engage him at this point – contrasted with Pompilia's spontaneous and courageous attempt at violence – will be a cause of shame and humiliation, as he is ridiculed by his neighbors at home. If Guido had not called in the law but rather had taken the matter into his own hands and killed the two at the inn, he would have been forgiven: "I say, the world had praised the man. But no! / That were too plain, too straight, too simply just! / He hesitates, calls law forsooth to help" (II, 1503–5). Instead of taking "the old way trod when men were men! Guido preferred the new path" and became "stuck in a quagmire" until he decided to do the right thing (that is, he killed Pompilia and the Comparinis and thus "Revenged his own wrong like a gentleman"), but, even then, Caponsacchi is left alive: "where's the Canon's corpse?" (1518–26). Because he supports Guido and his (supposedly justified) killings, Half-Rome, though he notes Guido's shame and humiliation after Castelnuovo, interprets his initial failure to act as a mistake in judgment rather than the result of cowardice, but other speakers will explicitly or implicitly contrast his later cold-blooded murders, assisted by young and brutal accomplices, with his (cowardly) failure to fight with Caponsacchi at the inn.

The Other Half-Rome interprets these key scenes from Pompilia's point of view; in her imagined narrative she sees:

> the man you call my husband? ay –
> Count Guido once more between heaven and me,
> For there my heaven stood, my salvation, yes –
> That Caponsacchi all my heaven of help,
> Helpless himself, held prisoner ...
> ...
> I sprang up, reached him with one bound, and seized
> The sword of the felon, trembling at his side,
> Fit creature of a coward, unsheathed the thing
> And would have pinned him through the poison-bag
> To the wall and left him there to palpitate
> As you serve scorpions, but men interposed –
> (III, 1153–67)

Pompilia's inspired action is contrasted with Guido's lack of nerve. Then the speaker recounts Guido's meeting with Caponsacchi "in secular costume / Complete from head to heel, with sword at side" (a description that contrasts with Half-Rome's satirical description of the priest as heroic lover) and "Guido, the valorous, had met his match, / Was forced to demand help instead of fight" (III, 1253–62). Then Caponsacchi was placed under arrest, and the speaker returns to the image of Pompilia, whom Guido assumes to be waiting helplessly in the room in bed, "writhing like a worm," but "this worm turned" and Pompilia "would have slain [Guido] on the spot / With his own weapon, but they seized her hands" (III, 1284–5), and the speaker has her address Guido as "The ignoble noble, the unmanly man, / The beast below the beast in brutishness!" (III, 1292–3). Among later, similar comments in his monologue, The Other Half-Rome in summary refers to "the wife's courage and cunning, – the priest's show / Of chivalry and adroitness, – last not least, / The husband – how he ne'er showed teeth at all, /

Whose bark had promised biting; but just sneaked / Back to his kennel, tail 'twixt legs as 't were.." (III, 1451–5). At this point, it is already clear that Guido's critics hold him in contempt for his cowardly ineptitude rather than simply his misuse of power and position as an oppressive and domineering husband.

Tertium Quid, whose lofty point of view is not directly tied to either side in the controversy but who is usually interpreted as leaning toward Guido's side because of his noble birth, also engages the issue of Guido's apparent cowardice and has Caponsacchi taunting Guido: "You, a born coward, try a coward's arms, / Trick and chicane,–and only when these fail / Does violence follow"(IV, 1087–9). Tertium Quid acknowledges that some ridicule Guido for his later "irrepressible wrath at honour's wound" after he has passed up his chance to strike out at the priest –"There's man to man, – nature must have her way, – / We look he should have cleared things on the spot" (IV, 1126–7) and comments that if he had slain the lover and the wife (or rather slain him and "since she was a woman and his wife" merely shamed her by stripping her naked) "Good! One had recognized the power o' the pulse" (IV, 1139). But after missing his opportunity when the noncombatant priest was "metamorphosed into knight"and turning to law, "let us hear no syllable o' the rage" (IV, 1144). He goes on to say that some of Guido's supporters account for this apparent cowardice by referring to his state of mind (Guido claimed that Pompilia had drugged him on the night before her flight) and the confused circumstances at the inn. And, even if Guido acted in a cowardly way, the institutions of the Church and the civil law supported his cowardice. In committing the murders, Guido finally went on to do what he should have done earlier, and those who insist that he should have acted sooner (or not at all) do not understand that a wound to the soul (unlike a wound to the flesh) "rankles worse and worse" with time (IV, 1524).

The reader is thus prepared for Guido's defense against the charge of cowardice in his first monologue, which comes next, in book 6. Browning has Guido speak in an antechamber of the same courtroom in which, eight months earlier, he had testified in the trial of Pompilia and Caponsacchi after their arrest in Castelnuovo. The three judges hearing Guido's case are the same who heard the earlier one. After describing his meeting with the now-familiar figure of Caponsacchi "[i]n cape and sword a cavalier confessed" in the courtyard of the inn, he quotes the now familiar argument "What follows next? / 'Why, that then was the time,' you interpose,'/ 'Or then or never, while the fact was fresh,' / To take the natural vengeance" (V, 1063–6). Guido's friends, as it were, cry "Kill!" at that moment, but when he does finally kill, after the "crime" of Pompilia and Caponsacchi, "only surmised before, / Is patent, proved indisputably" they cry "'So rash? ... / 'so little reverence for the law?'" (V, 1072–81). When Guido "called in law to act and help" the first time, his critics "cry / 'You shrank from gallant readiness and risk, / Were coward'" (V, 1085–6). Even if he were a coward, Guido argues, "Does that deprive me of my right as a lamb / And give my fleece and flesh to the first wolf? / Are eunuchs, women, children, shieldless quite / Against attack their own timidity tempts?" (V, 1092–5). Browning makes Guido's attempt to identify himself as a helpless victim ludicrous, but the use of gender here is significant. Men are expected to strike out in their own defense rather than appeal to law. Of course, Guido goes on to claim that his respect for law and the sanction of authority was his real reason for restraint and compares his humiliating position after Pompilia attempts to attack him with his own sword to that of a "tyrant" in a melodramatic "playhouse scene." He describes the "lovers'" absurd claim that the incriminating love letters found in their room were forgeries.[12] In his account of the inadequate legal measures taken against Pompilia and Caponsacchi after Castelnuovo, he employs phrases such as "I played the man as best I might." When he heard of Pompilia's pregnancy and her move from the convent to her parents' home, he describes his responsive rage in terms that are meant to re-create the violent passion that his critics assumed he should have felt toward Castelnuovo. When he confronts the hated Violante on the night of the murders, he is carried away by an "impulse, one / Immeasurable everlasting wave of a need / To abolish that detested life" (V, 1653–5). In subsequent rhetorical flourishes, Guido describes himself as carrying out the intent of the law and God's will.

In studying Browning's representation of Guido's questionable manhood, with an emphasis on his cowardice, it is important to consider the books in chronological order, to take note of the way Guido's character is developed; I have emphasized the process of accumulating details about Guido in the narratives. Browning places Caponsacchi's monologue, book 7, next for maximum contrast with Guido's, and the juxtaposition of the priest's not always polite but apparently

---

[12] As discussed below, Browning assumed that Pompilia was telling the truth when she claimed she was illiterate, but there is strong evidence in the OYB that she could read and write. A. K. Cook summarizes this evidence in Appendix IV of his *Commentary*; however, in his edition of the OYB, Gest goes much further in pointing out Browning's deviations from his source. In addition to the issue of Pompilia's literacy, Gest notes Browning's misunderstanding of many of the legal issues involved in the case (in Gest's view Browning unfairly ridicules the lawyers) and his apparently intentional misrepresentation of Caponsacchi, a subdeacon and Canon in the Church, as a priest. According to Gest, "He was not a priest, did not claim to be a priest, and is never called a priest in the Old Yellow Book," but Browning's characterization of him as such is followed "by all the Pompilia and Caponsacchi worshippers, whose adulation of him, as a soldier-priest, a saint, a Christian hero and a St. George is nauseating to any one who takes the trouble to read the cold facts of the story in the Old Yellow Book" (9). Furthermore, Gest judges that, based on the evidence given in the OYB, "it seems quite clearly established that Pompilia was guilty of adultery with Caponsacchi" (610), and if Guido had killed Pomilia and Caponsacchi at Castelnuovo, he probably would have been mildly punished or acquitted, but that, given the circumstances surrounding the murder of Pompilia and the Comparinis, his conviction and subsequent execution were to be expected. See also Beatrice Corrigan, *Curious Annals: New Documents Relating to Browning's Roman Murder Mystery* (1956). For Tucker, Browning's "getting it wrong" in this sense, the "failure of a minor premise that we can now ascribe to sentimental or Romantic idealization on the poet's part leaves intact his major historicist premise ... which makes *The Ring and the Book* not just a versified sensation novel but a nineteenth-century epic of the first magnitude" ! (*Epic*, 441).

spontaneous commentary and emotional appeals to God and principles of honor with Guido's blatantly artificial rhetorical ploys is of course central to Browning's construction of masculinity in the poem as a whole. As for the scene in the courtyard at Castelnuovo, Caponsacchi quotes Guido's insults delivered in a partially howling, partially hissing voice but observes "oh, how he kept / Well out o' the way, at arm's length and to spare!" (VI, 1415–16). In this version, as Guido addresses the authorities who have come to arrest Caponsacchi and Pompilia and remarks that he has refrained from drawing his own sword because Caponsacchi is a priest, the priest feels an urge to leap on Guido and strangle him with his hands: "one quick spring, / One great good satisfying gripe, and lo! / There had he lain abolished with his lie, / Creation purged o' the miscreate, man redeemed, /A spittle wiped off from the face of God!" (VI, 1450–54). His "poor excuse" for not acting at this key moment is that his initial reaction to Guido's ludicrous charges against Pompilia was mirth – because his lies were so absurd – and then the moment of opportunity was gone. Himself restrained by arresting officers, the priest watches helplessly as others prevent Pompilia from striking Guido with his own sword: "'Die,' cried she, 'devil, in God's name!'/ Ah, but they all closed round her, twelve to one, / – The unmanly men, no woman-mother made, / Spawned somehow! Dead-white and disarmed she lay. / No matter for the sword, her word sufficed / To spike the coward through and through" (VI, 1520–25). And Caponsacchi's continued angry denunciations of Guido through the rest of his monologue focus on his cowardice.

The passion of Caponsacchi's chivalrous narrative is contrasted with the quiet spirituality of the dying Pompilia, whose monologue follows in the seventh book. For Pompilia the evil Guido and her miserable life with him have become like a bad dream, which she would like to erase from her memory as she prepares herself for the afterworld – her already strong religious faith having been reinforced by a visit from her confessor, the priest Don Celestine – but she gives a spontaneous and straightforward account of her childhood filled with happy fantasies, her miserable life with Guido, her pregnancy, and the incidents that led to Guido's fatal attack. When she is first informed by Violante about the arranged marriage, she expects a "cavalier" but instead is introduced to "Guido Franceschini, – old / And nothing like so tall as I myself, / Hook-nosed and yellow in a bush of beard" (VII, 390–92). The impropriety of this May–December marriage, a point to which I will return, is reinforced by Guido's disgusting appearance and personality, and Pompilia must not only endure sexual relations with the husband she cannot love – "Why in God's name, for Guido's soul's own sake / Imperilled by polluting mine, – I say, / I did resist; would I have overcome!" (VII, 779–81) – but also the unwanted advances of Guido's brother Canon Girolamo, an "idle young priest," who "taught me what depraved and misnamed love / Means" (VII, 804–5). Pompilia's instinctive piety and religious faith, which never wavers in spite of her desperate situation, is rebuffed by the institutional church, as her appeal to the Archbishop for help is rejected, but eventually rewarded by the coming of the chivalrous young priest Caponsacchi, who instinctively understands her plight and takes on her rescue as a holy mission. In her narrative the cowardice of her husband is again stressed as he plots to create the appearance of his wife's infidelity with the aid of his maid Margherita (including forging "love" letters to Caponsacchi, which the priest understands at once cannot be genuine) and makes empty threats while brandishing his sword. In her account of the incident at Castelnuovo, after seeing her "angel helplessly held back / By guards" (VII, 1571–2), her thwarted attempt to strike the cowardly Guido with his own sword was "on impulse to serve God" rather than to save herself or her unborn child (VII, 1584).

In his rage Caponsacchi, whose urge to kill Guido is rendered problematic by his priestly office, has appropriated Pompilia's earlier moment of violent resistance and in his imagination identifies with her thwarted attempt to attack Guido, and his language is very similar to that of Browning's Count Gismond in the poem by that name – recalled by Gismond's wife – as he kills the man who has wronged and defamed her but forces him to retract his *lie* before he dies at her feet. Guido's *lie* is of course his charge of adultery, and Caponsacchi in his angry passion and Pompilia in her adamant sincerity give the lie to Guido. Caponsacchi's claim of chastity in the context of what he (and Browning) must understand to have the appearance of an archetypal narrative of two men fighting for the sexual possession of a woman depends on the construction of his own role as that of chivalrous knight, reinforced by his spontaneous, instinctive association of Pompilia with the Madonna. In her monologue, Pompilia explicitly accepts this association, and like Marian in Barrett Browning's *Aurora Leigh*, she acknowledges no human father for her son: "No father that he ever knew at all, / Nor ever had – no, never had, I say!" (VII, 90–91). Pompilia's attitude is informed by her own experience, for she herself, the child of a prostitute, is fundamentally fatherless: "My father, – he was no one, any one ... he who came, / Was wicked for his pleasure, went his way, / And left no trace to track by" (VII, 290–93).

Thus the saintly Pompilia, while offering an absolute denial of adultery, reinforces the image of a hateful, cowardly Guido not only in her intimate personal description of his abusive treatment of her but by cancelling his claim to fatherhood, a role that would implicitly support his vengeful actions, however questionable and problematic in some ways, in the service of his family's "honor." The monologues of the lawyers intervene between the testimonials of Caponsacchi and Pompilia and that of the Pope, offering "breathing space" and even some comic relief (in Browning's implicit satire of these men and their profession) before the Pope's very serious final judgment that will in his account absolutely verify the essential truth of the heroic protagonists and the essential falsehood of the villainous Guido. Guido's lawyer Archangelis bases Guido's defense solidly on the principle of honor and justifiable revenge – "Honour is man's supreme good" (VIII, 577), especially for men of Guido's privileged social position – and he bends legal technicalities and Christian morality to accommodate this idea, which is implicitly based on natural law. Of course Archangelis, like the other supporters of Guido, must answer the key "legitimate" question that might be asked by reasonable men – "'Why procrastinate'" (969) why did Guido not act on his (rightful and honorable) sense of outrage at Castelnuovo?

> "Right, promptly done, is twice right: right delayed
> Turns wrong. We grant you should have killed your wife,
> But o' the moment, at the meeting her
> In company with the priest: then did the tongue
> O' the Brazen Head give license, 'Time is now!'
> Wait to make mind up? 'Time is past' it peals." (VIII, 971–6)

In formulating his own answer to the charge that Guido had thus "forfeited [his] chance" (VIII, 981), Archangelis begins with Guido's own justification, that the soul, "where honour sits and rules," is unlike the body, which feels the "smart" of a wound "worst at first": "Longer the sufferance, stronger grows the pain" (VIII, 989–90). Then he goes on at length to prepare his arguments, dissecting the law and quibbling with the charges against Guido and his use of the four accomplices, who must be innocent if Guido is, and, after all, Guido did not "desecrate" his deed, did not "vulgarize vengeance" by actually paying them the money he promised them (there is evidence that they intended to kill Guido for this reason): "He spared them the pollution of the pay" (VIII, 1610). Repeatedly he returns to the issue of honor and righteous, justifiable vengeance: "Is the end lawful? It allows the means" (VIII, 1463). Literally, the end justifies the means. Guido's apparently questionable methods finally are irrelevant because killing his wife and her "clan" was motivated by his wounded sense of honor. Archangelis is not only answering the official charge of murder but deflecting the unofficial but potent and well-understood charge of Guido's cowardice and flawed manhood.

Not surprisingly, Pompilia's lawyer Bottini, in preparing his side of the case, emphasizes Guido's shameful, unmanly behavior toward Pompilia, for example, "[bursting] upon her chambered privacy" (IX, 850) and advertising her supposed adultery: "O thou fool, / Count Guido Franceschini, what were gained / By publishing thy shame thus to the world?" (IX, 870–72), and of course he stresses the incident at Castelnuovo. In striking contrast to Guido's cowardice, she drew his own sword "since swords are meant to draw" (IX, 892). As already implied above, Browning's portrayal of Bottini ensures that, in spite of his defense of Pompilia, he does not share with Caponsacchi and the Pope positive associations with her innocence and goodness. His unsympathetic method of first assuming the charges of adultery are true before deconstructing them is exacerbated by the fact that his language is saturated with debasing, insulting references to women's sexuality. Buckler shows this in considerable detail and argues: "To poison the world against women is such an obsession with him that he provides a virtual handbook of antifeminist techniques, a guide to female defamation" (204). Buckler's description of him as an extremely sexist, misogynous bachelor is perhaps overdone – "Like evil, he is beyond rational understanding ... [S]exism ... is a moral cancer in any culture" (207) – but there is little doubt that Browning's representation of Pompilia's advocate as a man unsympathetic to women is one of the profound ironies in the poem and is consistent not only with the implied author's negative views of lawyers and the legal profession but also with his questioning of masculinity throughout the poem.

The Pope[13] is often considered to be Browning's spokesman, despite his own limitations and uncertainties and despite his keen awareness of the multifariousness and mutability of the world and the fallibility of his Church and its representatives, because he amplifies and reinforces the judgments that Guido is intended to be seen as corrupt and evil and that Pompilia is to be seen as pure and good, moral assumptions apparently intended by the implied author. In the final analysis, Browning's Pope makes his decision to deny Guido's pardon (thus assuring his eminent execution) intuitively: he will "call good good / And evil evil" (X, 1873–4), valorizing the fundamental insight of the technically improper but heroically chivalrous priest and his attempt to rescue Pompilia. The aged Pope, who feels that he is near death, fears the future consequences of Enlightenment rationality and the demise of Christian heroism. Comparing himself to Guido, man to man as it were, he notes that despite his advanced age, "to-day / Is Guido's last: my term is yet to run" (X, 335–6). To him Guido is an animal who sinks lower than any other: "Not one permissible impulse moves the man" (X, 536); but in addition to his charges of bestiality and lack of spirituality, the Pope, like all of the previous critics of Guido, turns to gender. He notes that Guido's violence is "made safe and sure by craft" (X, 591), "Always subordinating (note the point!) / Revenge, the manlier sin, to interest / The meaner ..." (X, 598–9). The Pope goes on to condemn Guido's manhood in terms very similar to those used by his other critics. His forging of the love letters was "Unmanly simulation of a sin" (X, 647). After praising the actions of "the priest and wife" at Castelnuovo, the Pope gives his version of Guido's confrontation, "face-to-face" with Caponsacchi: "Who fights, who fears now? / There quails Count Guido, armed to the chattering teeth, / Cowers at the steadfast eye and quiet word / O' the Canon at the Pieve! There skulks crime / Behind law called in to back cowardice!/ While out of the poor trampled worm the wife, / Springs up a serpent!" (692–8). The Pope's choice of words is interesting. Pompilia as "serpent" here is not, cannot be, associated in any way with the Devil or with evil – she is after all pronounced by the Pope to be "First of the first" (999), "Perfect in whiteness" (1001). The "serpent" metaphor here stands for righteous, justified violence, and the reader may recall Pompilia's earlier reference to an "old rhyme" about a virgin, faithful to God, who was hiding herself in a cave to escape from pursuing "Paynims"; and, when they found her, thanks to an illuminating flash of lightning, she appealed to "God's fire" as well and was provided a sword with which she slew all of her attackers and "o'er the prostate bodies, sworded, safe, / She walked forth to the solitudes and Christ" (VII, 1388–9).

---

[13] Ostensibly, Innocent XII (Antonio Pignatelli, 1615–1700), but, as many scholars have pointed out, the views of Browning's Pope are much closer to the poet's own than to those of the historical Catholic Pope. Browning saw parallels between the late seventeenth century, the eve of the Enlightenment, and his own age of transition and doubt, and he created a character to draw them out. At the same time, however, it is misleading to interpret the Pope's opinions and interpretive methods as being synonymous with those of Browning.

Among additional references to Guido's flawed masculinity in the Pope's monologue is an observation that Guido took "unmanly means" to reach his end. Man "[m]ay sin, but nowise needs shame manhood so" (X, 718). So even for the Pope, Guido's guilt is exacerbated by his failure to live up to the standards of true manhood. The Pope also focuses on Guido's unnatural, cynical feelings toward his "first born, his son and heir" (X, 750) – who he thinks will now entitle him to inherit the Comparinis' property though their revelation of Pompilia's illegitimacy had cancelled her dowry. Guido's substitution of self-interest for fatherly pride and affection is an affront to both Christian morality and selfless fatherhood as defined by standards of masculinity.

Even if the reader does not take the Pope as Browning's spokesman or accept his judgment that Guido is essentially evil, Guido's second monologue, which follows the Pope's unequivocal condemnation, is not calculated to elicit the reader's sympathy. The Pope's certainty about his own judgments in spite of his questioning of his institutional "infallibility," his assumption that Guido is immoral but of sound mind, and other factors open up subtle questions about the old man's thinking characteristic of Browning's method but the implied author's condemnation of the murderer Guido is reinforced, and Guido's monologue represents perhaps the most extreme example of Browning's attempt to "get inside the mind" of a character who is alienated from the values implicit in the work in which he appears. Unlike, for example, the Duke's speech in the short monologue "My Last Duchess," Guido's monologue has been elaborately framed by thousands of lines of poetry, spoken by himself as well the other characters with various perspectives, and I have shown how, for example, aspects of his masculinity have been developed in certain consistent ways through the various monologues.

Guido is speaking, after the death of Pompilia and shortly after the Pope has rejected his final appeal, to Cardinal Acciaiuoli and Abate Panciatichi, both men of the Church who have had ties to Guido and his family and who might be seen as constituting a relatively "sympathetic" audience, given Guido's hopeless position. Instead of cooperating with their mission on behalf of the Pope to hear his final confession and encourage him to beg for forgiveness and take this last opportunity to save his soul, Guido rants on compulsively, often addressing the two churchmen directly and individually and occasionally pleading for his life but not encouraging response or dialogue. Early in his monologue, Guido says that it had been his folly to accept the "saw" that "'A man requires a woman and a wife'" (XI, 161) and ironically implies that his current desire to talk and reveal himself makes him feminine, referring to "this unmanly appetite for truth" (XI, 170). So deeply flawed is Guido's concept of masculinity that truth-telling in itself – as opposed to a rhetorical argument for one's own advantage – is assumed to be feminine, and a man does not have a fundamental need to mate with a member of the opposite sex. Despite references to Guido's (perhaps sadistic) sexual exploitation of his unwilling wife, there is no evidence that he is motivated by a strong male sexuality, and this is consistent with Guido's failure to express in any sense a sincere love for either his wife or newborn son. Browning portrays a Guido *driven* to reveal himself against his conscious will and of course compelled to express himself according to the inner logic of his mind.

Browning constructs this book so that it is as if Guido speaks continuously from the time the two enter his cell in the early morning (after he has been awakened with the news that he will die at sunset) until the guards who will accompany him to the place of execution arrive at the end of the day. Instead of implicitly revealing his true nature behind elaborate (though ineffective) rhetoric as in book five, Guido not only openly accepts his role of "wolf" to Pompilia's "lamb" but, while alternating between a defiant and wheedling tone, revels in his identity as a cynical, self-centered, amoral pagan. His clear assumption is that everyone who professes to believe in the conventional beliefs and moral values of the Church and society is a hypocrite. Instead, he embraces his *natural* privileges as a man – it is "God's law" that men control women through brute force, for example – and he assumes that his (inherited) rank and social status are his ultimate source of truth and value. Beyond that, he asserts that the Aretine – and ancient Etruscan – cultural traditions associated with his lineage have preserved these privileges to a far greater extent than have Rome and the Church as an institution. He relates how he was surprised and disgusted by Pomplila's piety and meekness. On the one hand, he argues that Pompilia, as his wife, had the duty to feign affection for him, but, on the other, that it was unsettling and unnatural of her to obey her mother and accept his rough treatment without complaint.

He describes the central incidents at Castelnuovo as bad luck:

> Had things gone well
> At the wayside inn: had I surprised asleep
> The runaways, as was so probable,
> And pinned them each to other partridge-wise,
> Through back and breast to breast and back, then bade
> Bystanders witness if the spit, my sword,
> Were loaded with unlawful game for once –
> Would you have interposed to damp the glow
> Applauding me on every husband's cheek?
> Would you have checked the cry "A judgment, see!
> A warning, note! Be henceforth chaste, ye wives,
> Nor stray beyond your proper precinct, priests!" (XI, 1534–45)

In contrast, luck was on his side when he and his accomplices attacked the Comparinis – up to a point. It is as if the three victims were waiting there to be murdered (see XI, 1575–605), but later the watchman at the gate unexpectedly refused to be bribed (as is frequently pointed out in the various narratives, Guido had made the absurd mistake of not obtaining the simple pass that would have allowed him to return through the Roman city gates), and he and his men were not able to re-enter their Tuscan homeland. Guido's account of the fabricated story he would have told, had he been able to escape from the scene of the crime, is revealing:

> I should have told a tale brooked no reply:
> You scarcely will suppose me found at fault
> With that advantage! 'What brings me to Rome?
> Necessity to claim and take my wife:
> Better, to claim and take my new-born babe, –
> Strong in paternity a fortnight old,
> When 'tis at strongest: warily I work,
> Knowing the machinations of my foe;
> I have companionship and use the night:
> I seek my wife and child, – I find – no child
> But wife, in the embraces of that priest
> Who caused her to elope from me. These two,
> Backed by the pander-pair who watch the while,
> Spring on me like so many tiger-cats,
> Glad of the chance to end the intruder. I –
> What should I do but stand on my defence,
> Strike right, strike left, strike thick and threefold, slay,
> Not all–because the coward priest escapes.
> Last, I escape, in fear of evil tongues,
> And having had my taste of Roman law.' (XI, 1700–19)

Guido's fantasies about these key events in the poem's narrative – probably as spontaneous and genuine as Caponsacchi's emotional account of his failed quest to save Pompilia and Pompilia's own account of her own single-minded mission, inspired by her pregnancy, to escape from Guido – provide the reader with the most comprehensive understanding of how Guido's failed masculinity is the key to his failure as a human being. At Castelnuovo, if only Guido could have skewered the sleeping bodies of Pompilia and Caponsacchi with his sword. No confrontations with the warrior–priest or the fearless runaway spouse, just the opportunity to kill them in their sleep – in a way that signals not only justifiable violence but a righteous and admirable display of patriarchal authority. Although that opportunity did not materialize, a successful escape after the murders would have allowed him to create another version of a justified attack on the adulterous couple, this time including an opportunity to shame the supposedly chivalrous priest as in fact a coward who fails to defend Pompilia and runs from the righteously violent husband to save his own skin. Guido's pathetic desire to shift the *appearance* of his own patent cowardice to the priest highlights the emptiness of his self-image.

Apparently, Guido cares only about his own self-interest, does not distinguish between truth and falsehood in his statements to others, and does not believe in the Christian God. But just as he is shown to be bereft of moral and spiritual values, he is also bereft of the positive masculine qualities that would supposedly be associated with even the *pagan*, "natural" order that he opposes to that of the hypocritical Christians. He thinks of his son only in terms of using him for his own self-interest and the "paternal" role he would feign for the benefit of society is a fabrication – there is no sign of a genuine pride in his roles as husband, father, or, in a larger sense, patriarchal authority to be passed down to future generations.

Critics who point out Browning's representation of "sexism" in *The Ring and the Book* are, of course, correct in sensing the poet's implicit criticism of a traditional society in which men dominate and control women. However, the whining, cowardly Guido does not represent traditional masculine values of strength and authority. He does not offer physical force and courage as an alternative to a potentially weak and effeminate spirituality. On the contrary, Browning, through his own commentary as the poet–narrator and his development of Caponsacchi and the Pope as positive (if not idealized) characters, attaches the values of chivalry, not to the institutionalized church, but to his concept of a kind of evangelical Christianity based on an individualized, intuitive access to God's truth. Browning could not afford to formulate a Guido who resists Christian ideals by appealing to natural instincts, that is, to the "fallen man" as formulated by St. Paul or the Darwinian nature "red in tooth and claw" as formulated by Tennyson.

Caponsacchi, in spite of his ambiguous social status as a chivalrous priest, is more natural and "authentic" than Guido. Frustrated by his superficial role in the church, acting as a kind of public relations man – promoting the church among prominent patrons and especially popular with fashionable ladies – he suddenly discovers a truly "spiritual" mission (as verified by the Pope himself) when he spontaneously responds to the appeal in the "captive" Pompilia's visage. The fact that Browning closely associates his narrative with the Andromeda myth and its "Christian cognate," the legend of St. George and the Dragon,[14] has been widely discussed over the years. As William C. DeVane pointed out in a 1947 article:

> In those books ... where the speakers give favorable judgments upon Pompilia and Caponsacchi, I have counted at least thirty references to the Andromeda and its cognate myth, not counting such facts as this–that Browning, for all his accuracy and care in consulting the Astronomer Royal upon the condition of the moon on the night of Pompilia's flight, April 29–30, 1697 ... changed the date ... so that the flight would fall on April 23, St. George's Day.... [W]henever Browning is representing, favorably to Pompilia and Caponsacchi – and that is a great deal of the time – the great scene at the inn at Castelnuovo where the real conflict between the opposing forces takes place, he habitually and consistently thinks of it in the terms of the Andromeda situation, with Caponsacchi as Perseus, Pompilia as the manacled victim, and Guido as the dragon. (42)

In this sense *The Ring and the Book* is far from being unique in Browning's oeuvre. DeVane points out that Browning frequently uses the Andromeda myth throughout his career to express his personal version of religious faith in a threatening world dominated by the ideology of scientific rationalism (36). Later critics, such as

---

[14] According to an ancient Greek myth, Andromeda was the daughter of Cepheus and Cassiopeia, living on the shores of Asia Minor. To appease Poseidon, the god of the sea, who had been offended by Andromeda's father, she was chained to a rock in order to be sacrificed to a dragon. The Athenian hero Perseus slew the dragon, set Andromeda free, and married her. The legend was later adapted and applied to the third-century Christian hero St. George, patron saint of England, Aragon, and Portugal.

Langbaum (1966) and, most extensively, Adrienne Munich (1989), also deal with Browning's use of the myth. Munich, who examines the widespread treatment of the myth by various male Victorian writers from a feminist perspective and discovers "envy, fear, curiosity, unease, as well as a conservative effort to classify women as docile Andromedas and men as stalwart Perseuses" (3), observes that frequent allusions to Andromeda and the other mythological characters in *The Ring and the Book* "represent absolute moral positions, counterbalancing the relativism of the differing viewpoints of the poem's monologuists" (140). Browning, who inherits both the Evangelical Christian's desire for intuitive faith (principally from his mother)[15] and the Romantic poet's displaced religious desire to locate moral universals in Nature from Shelley and other key literary influences, developed a highly gendered concept of "God's truth" insofar as it can be expressed in literary art. His intuitive belief in a transcendent truth is framed by a cultural consensus that has thoroughly absorbed and normalized basic ideals about the goodness and purity that can be located in uncorrupted women, women who have the capacity to spiritually save potentially evil males who willingly (and perhaps inspired by God) adopt a chivalric code, and this use of chivalry can be compared to Tennyson's in the *Idylls*, as discussed in Chapter 2. In *The Ring and the Book*, as DeVane puts it, "we see Pompilia-Mrs. Browning[16] – Andromeda rescued from the dragon Guido by Caponsacchi-Browning-Perseus, first; and then later, when truth or justice is endangered, Pope Innocent, the Vicegerent of God, is the rescuer" (37). In this formulation, the "dragon Guido," quite apart from Browning's "personal myth," corresponds to Victorian gender stereotypes associated with the nineteenth-century reconstruction of medieval "chivalry" and related cultural formulations discussed in Chapter 2. However, Guido as a dragon, or even as a wolf, could potentially represent a violent, hypermasculine figure. Instead, as I have shown, he is a cowardly, "unmanly" man, empty of not only moral values but masculine standards of conduct, who whines and plots rather than standing up for his position. At the same time, as a dysfunctional and abusive husband and father, he also lacks any traces of the alternative Victorian "domestic manhood" described in Chapter 1. Furthermore, his failure to advance in his career within the Church demonstrates his deficiencies within the context of male hierarchies. Browning is of course critical of the institutional Church, but Guido's failure to "fit in" and prosper is represented in a negative way, in contrast to Caponsacchi's unsettled position.

[15] The central importance of Browning's relationship with his mother and her Evangelical Christian faith was first pointed out by Betty Miller in her biography.

[16] Browning's biographers and critics often point out the apparent association of the character Pompilia with his idealized memory of his wife, Elizabeth Barrett Browning. For example, Altick and Loucks write, "The story enabled [Browning] to treat at great length the theme of a beleaguered woman and her chivalrous rescuer, and to relive, in a way, his own glorious adventure in 1846 as a Perseus-St. George. While there is little overt resemblance between Pompilia and Elizabeth Barrett … it is likely that Pompilia is in some ways a much idealized version of Browning's dead wife, or perhaps more accurately, a substitute figure" (19–20).

In secular terms, Guido is a failure as a "count" as well, and rather than advancing his family's fortune and status, he completes the historical process of its decline. According to the OYB, Guido was 36 at the time of his marriage to Pompilia, 40 at the time of his trial and execution. Browning adds 10 years to his age. This change of course exaggerates the already pathetic distance between him and his child bride, who is only 13 at the time of her marriage, 17 at the time of her death. Pompilia's role as victim is implicit in the OYB, but Browning's change of Guido's age is consistent with other changes, such as his representation of her illiteracy as a fact that is uncontested by the implied author: the lying Guido insists on the validity of the forged love letters, but in a way that only reinforces their inauthenticity. However, there is evidence in the OYB that Pompilia could in fact read and write.[17] On one level, Browning's change protects Pompilia's saintly innocence while simultaneously reinforcing not only the guilt and fundamental falsehood of Guido but also his lack of masculine pride and willingness to expose his family name to public shame.

Another significant change is in the birth order of Guido and his brothers. Browning portrays the Abate Paolo as Guido's younger brother. As a second son, Paolo pursued a career in the Church with some success, and in Browning's poem (but not in the OYB), he negotiated on behalf of his brother with Violante in arranging for the marriage with Pompilia. Canon Giralamo, a still younger brother and also a priest, is accused by Pompilia in her monologue of making improper sexual advances. However, the historical Guido was not the eldest son, as Browning portrays him: Paolo, the eldest, was seven years older than Guido and Girolamo was four years older.[18] Browning's Guido is the patriarchal head of the Franceschini family, and his intense sense of privilege associated with that position is one of the chief targets of Browning's criticism in the poem.

One of the most insidious actions taken by Guido as family patriarch is of course his recruitment of the accomplices who assisted him in the murders: four workers employed on his estate. Guido refers to them as "Resolute youngsters with the heart still fresh"(V, 1555), and according to evidence in the OYB, they were probably all in their 20s; in the trial, special claims were made for one of them, Francesco Pasquini, because he was technically a "minor," under the legal age of 25.[19] Domenico Gambassini also claimed to be a minor, though his claim was not substantiated, and apparently he was not directly involved in the killings. In Browning's poem, the four are treated almost wholly as appendages of Guido. The contrast between their execution by hanging prior to Guido's beheading at the guillotine symbolizes the gulf between their social rank and that of Guido. It

---

[17] See note 12 above.

[18] See the reference to Beatrice Corrigan's edition of related documents in the notes to *The Complete Works of Robert Browning*, vol. 7, 291. Apparently there was also a fourth son, Antonio Maria, younger than Guido, but he is mentioned by neither Browning nor the OYB.

[19] See Gest's annotations to his edition of the OYB, 516–17.

is interesting that the poem does nothing to question this social distinction. With his life in the balance, Guido delivers thousands of lines of poetry and though he may be said to justify in full measure the taking of his life, the question of his mortality is continuously at issue, while the four accomplices remain unexamined as individual human beings. I am not implying that the evidence of the OYB suggests they are innocent – and their apparent intention to kill Guido himself after he fails to pay them their "blood money" in Browning's version of the story is not calculated to elicit sympathy from the reader – but it is as though their individual lives, unlike Guido's, are not at issue.

The Pope refers to the four as both "God-abandoned wretched lumps of life" (X, 922) and "stout tall bright-eyed and black-haired boys" (X, 929). He asks, "Are these i' the mood to murder, hardly loosed / From healthy autumn-finish of ploughed glebe, / Grapes in the barrel, work at happy end, / And winter come with rest and Christmas play? / How greet they Guido with his final task – / As if he but proposed 'One vineyard more / To dig, ere frost come, then relax indeed!')" (X, 932–8). The Pope assumes they acted purely out of greed and did their deed "purely for the pay," setting aside the idea that Guido's "feudal tenure claims its slaves again" (X, 946). However, even if the young men were motivated by selfish greed rather than a mindless loyalty and obligation to their "feudal" lord, Guido's "raiding party" constitutes an ancient tradition indeed, one that is associated with countless pre-state societies, whereby young men commit acts of violence in the service of their chief, justified by issues involving "honor" and vengeance. The fact that their victims are unarmed and two of them women of course heightens the sense of "uncivilized" terror and unforgivable criminality. As we have seen, Guido senses the potentially more favorable interpretation of these killings if he could have successfully fabricated a version in which Caponsacchi had been present but had run away from the fight rather than try to defend the two women and old man. As it is, the young men are assumed to be judged as their leader Guido is judged – if Guido deserves to die, they, as his henchmen, deserve to die, also, regardless of the individual acts actually committed by each one. Browning does not challenge this ancient masculine tradition.

Setting his narrative in (documented) historical time rather than Tennyson's mythic past, Browning nevertheless deals with issues of masculinity that are intimately tied to the universals of human nature, especially issues of male violence and male sexual and social roles, and in general his representation of masculinities tied to these universals, like Tennyson's, focuses on tensions, complications, uncertainties. His method is quite distinct from Tennyson's, however, in revealing the complexities of "manhood" and in some ways is more closely related to Barrett Browning's narrative with its "contemporary" setting. Munich argues that Browning "appropriates" the feminine "power" of Pompilia, and "as an emblematic device, Andromeda becomes a sign of his own poetic genius" (145). There is a profound irony in this formulation, for the supposedly illiterate Pompilia is not a poet, not a creative artist: she simply is herself, and she speaks directly from her soul, without the corrupting influence of "discourse" (the most extreme form of which is found in the legal documents created by the lawyers). Thus in this scheme the poet gives voice to her "truth," which at points in the narrative is associated with God's truth. There is some validity in Munich's claim, because, as many readers and critics have observed, Pompilia is at the heart of the poem's meaning: as Buckler puts it, her monologue "is the imaginative crux of *The Ring and the Book* ... the poem-within-a-poem by which the integrity of everything else must be finally decided" (162). The other characters are implicitly evaluated in terms of their relation to her. Most significantly, of course, Guido is judged as her persecutor and oppressor, while Caponsacchi and the Pope are her champions, and these fundamental relationships shape the poem's meaning. Guido, like Milton's Satan (though in no way sharing his "grandeur"), is established as a fascinating if despicable villain whose ultimate doom is assured. His disgusting but entertaining second monologue, with its interesting allusions to late seventeenth-century Tuscan and Roman culture, can be safely allowed to continue at any length because the voice of the poet, sustained by Pompilia's gentle perfection, will inevitably prevail.

Although Browning in this sense "appropriates" the feminine in a way that reminds us of the observations of previously discussed scholars who have linked Browning with a feminine point of view, his poetic voice is ultimately not feminine but rather a masculine one that affirms an idealized feminine as a key to fundamental human values, and one that judges *other* masculine voices in a relentless manner in terms of male–male competition. I have shown in considerable detail how Guido's masculinity is consistently analyzed and judged throughout the poem – his status as villain is not in question, and his belief that "manliness" is equivalent to selfish dominance is condemned, but his distortion of conventional masculine codes in his consistent and fundamental selfishness and cowardice is not the only significant issue involving masculinity. The ultra-feminine Pompilia attains her essential identity naturally, by accepting who she really is, by accepting her pregnancy not as a result of unwelcome or forced sex with her detested husband but rather as an almost mystical state associated with God's truth in the tradition of the Madonna. The sympathetic male characters, on the other hand, must construct precarious masculine identities by their considered actions: in accepting his role as warrior–priest, Caponsacchi discovers meaning in his life, but he must act it out and he is not entirely successful. As Buckler puts it, "What [Pompilia] did for him was to rescue him from the self-destruction upon which he had embarked and to lead him so far from the erosive ambiguity of an erotic priest vowed to celibacy as to give him a decent, representatively faulted man's chance to do his duty and save his soul" (157), and he is at the end "a heavily laden, deeply faulted, profoundly sympathetic representative of our race" (160). The complications in his character as described here are related to his ambiguous masculinity despite his heroic role playing.

Male celibacy is not only an important element in Caponsacchi's complex masculinity but a recurrent issue in the poem. Guido's younger brothers, both priests, are sworn to celibacy. Browning has Paul (Abate Paolo) act as a kind of surrogate for Guido in negotiating with the Comparinis for their daughter and her dowry, and later Girolamo, according to Pompilia, attempts to have sex with her. In a contrasting positive context, the patriarchal yet celibate Pope must act with masculine decision, most notably in his decision to end Guido's life. This is consistent with his validation of the warrior–priest's "healthy rage" (X, 1133) and his recognition of "healthy" masculine violence reminds us of his comment that revenge compared to "interest" is "the manlier sin" (X, 598). His masculinity is also affirmed in his "spiritual adoption of Pompilia as his daughter."[20]

Browning's "chivalrous Christianity" incorporates conventions of traditional masculinity, which valorize "healthy" violence in a righteous cause motivated by genuine idealism, and in *The Ring and the Book* the idealization of Pompilia provides the focus for a system of values that relentlessly interrogates the masculinity of the male characters. This is not to say the other principal women characters are represented as morally superior: Pompilia's biological mother is a prostitute who sells her child to Violante, who in turn manipulates her innocent young adoptive daughter mercilessly (though Pompilia has memories of a happy childhood) and in essence "sells" her to Guido. This is very similar to what Mirian's mother does in *Aurora Leigh*. However, the emphasis is overwhelmingly on Guido and his defective character, most emphatically on his defective masculinity. There is substantial evidence that Browning's presentation is consistent with larger cultural patterns of development. For example, Martin J. Wiener's *Men of Blood: Violence, Manliness and Criminal Justice in Victorian England* (2004) is an extensive study of the crimes of homicide and rape committed by men during the Victorian age, how these crimes were handled by the criminal justice system, and how a discourse about men's violence simultaneously developed in the popular press as well as in commentary by lawyers and judges involved in the system. Wiener finds two significant trends. First, there was an overall decline in the incidence of violent crime in general, along with an increasing social intolerance of violent and "disorderly" public behavior by men and an increasing emphasis on male self-discipline. Second, there was an evolving "reconstruction of gender" associated with an increasing emphasis on punishing male violence against women, who were "increasingly seen as both more moral and more vulnerable than hitherto" while men were seen as more dangerous. Wiener generalizes that during the long reign of Queen Victoria (1837–1901), "the treatment of women in Britain and in the burgeoning empire became a touchstone of civilization and national pride" (3).

This cultural shift was of course complex and uneven, and it was contested by those who championed traditional codes of honor valorizing male-on-male violence and those who would hide domestic violence from public scrutiny. Furthermore, British class consciousness played an important role in the increasing condemnation of the kinds of drunken violence and brutality associated with the lower classes. Nevertheless, the overall trends – across class lines – identified by Wiener are well documented. Many of the sources he mines for this book have never been used before in a study of this scope, and some archives have

---

[20] See Woolford and Karlin, 85.

become available to researchers only in recent years: "discussions between Home Secretaries, their civil servants, and judges, together with appeals from condemned prisoners and others for mercy" (xi). Much of the value of his project, however, is in his readings of newspaper and other published accounts of "sensational" crimes, which of course constitute a literary context for genres like the sensation novel, the melodrama, and "serious" literature like *The Ring and the Book*.[21]

Wiener's findings are consistent with other recent research that has called attention to the dramatic decline in male homicide rates in post-industrial European democracies, as discussed in Chapter 1. He then turns to his primary focus, gender-based violence, which is especially relevant to Browning's frame of values in his poem. Obviously, the history of "violence against women" has been discussed a great deal by feminists during the past few decades, but one of Wiener's major contributions is to show that in spite of widely held stereotypes about the "patriarchal" and "misogynist" nineteenth century, there was a marked increase in prosecutions and convictions for rape and sexual assault. Like all the major generalizations he makes about the Victorian era, this one is based on factual evidence, in this case 174 reports of rape trials at the Old Bailey in sampled years and all reports of English rape trials (about 800) appearing in the *Times* between 1790 and 1905.

In a chapter on the growing contrast between homicidal men and women, Wiener shows an increase in the level of punishment for wife-killing, and, despite a growing opposition to capital punishment that lowered the overall number of executions, "Even per capita, more wife-killers were being hanged at the close of Victorian's reign than in its early years" (165). Then he examines the persistent but nevertheless changing traditional attitudes toward "bad wives" who "deserved" to be punished by their husbands because of provocations such as drunkenness or for the sin of adultery. Increasingly the notion of acceptable "chastisement" of "bad wives" was dismissed from courtroom discourse, and tolerance for "crimes of passion" by the husbands of unfaithful wives declined. Wiener devotes an entire chapter to "Probing the Mind of a Wife Killer," taking a close look at defense arguments based on the killer's state of mind, and finds that in the limited area of "insanity and the related conditions of delirium tremens or epilepsy," it became "easier to negate the existence of evil intention," but, overall, "men who killed (and even more, who sexually assaulted) women were treated more severely in comparison to other offenders at the end of Victoria's reign than had been true at its start" (288). And although Browning obviously had a historical interest in late seventeenth-century Italian culture, the value frame of his poem's implied reader and implied author is that of nineteenth-century England.

---

[21] In my essay "The Male Villain as Domestic Tyrant in *Daniel Deronda*," I compare George Eliot's version of a "domestic tyrant" in her novel with that of Browning in *The Ring and the Book*, and I refer to Browning's "melodramatic imagination," using the term introduced by Peter Brooks in his book by the same name.

Wiener emphasizes the "changed conception of manliness at the heart of Victorian ideology" that helped to create a "peaceable kingdom" paradoxically "ruling a vast Empire that rested ultimately on force" (290). He points out another paradox when, referring to the Victorian legacy today, he observes that "contemporary feminism, for all its repudiation of Victorian values, continues – usually without acknowledgment – to draw upon that well for nourishment" (291). In placing Browning in this context, we must be careful not to ignore a certain ambivalence toward violence on his part: his "chivalrous Christianity" retains a role for the impulsive passion that may inspire men's courageous acts of violence against other men in a righteous cause.

Browning's critique of Victorian masculinity should also be seen in the context of his generally negative representation of family life in *The Ring and the Book*. As Woolford and Karlin point out, Guido "cannot conceive of his son as his own flesh and blood except in the sinister sense of a devouring parasite," and behind the apparent "home-loving respectability" of his lawyer Archangelis, who in a superficial way focuses on the welfare of his own son, "is a selfishness as predatory, soulless, and cynical as Guido's" (82). In the final analysis, "[t]he family in *The Ring and the Book* ... is a site of deceit, corruption, violence, and greed; marital, parental, and filial relationships are endorsed only when they are figurative, as in the 'marriage' of Pompilia and Caponsacchi.." (85). Woolford and Karlin go farther: "how typical a view is this in Browning's poetry as a whole? Thoroughly, is the answer"(85).

In the previous chapter, I point out Clough's emphasis in *Amours de Voyage* on cognitive activity and the protagonist Claude's persistent urge to analyze – in the context of his own life history – the "somatic" struggle for survival along with "reproductive" life goals associated with mating, parenting, kin relations, social relations. Browning's presentation of a literary narrative constructed from documents distanced by historical time and cultural difference incorporates dysfunctional applications of these behavioral patterns related to mating and socialization, while formulating a Romantic narrative based on a "figurative" relationship that ends in tragedy. But what of cognitive behavior in Browning's story? Our study of masculinity in *The Ring and the Book* helps us to understand the gap that has been traditionally perceived between Browning's "idealism" and his "pessimism." It is useful to contrast his representations of Guido and the Pope. The cognitive abilities of the Pope are focused on a comprehensive vision of humanity based on an understanding of history in terms of fundamental moral and spiritual values that transcend the theological conventions of the institution he heads. Guido's limited intelligence is fully focused on what he perceives to be his own petty self-interests, with a nearly total disregard for the welfare of others. As for "theory of mind," the Pope understands Guido's intimate thoughts and feelings, just as Browning and the reader do, but Guido has no inkling of a consciousness like that of the Pope. The Pope is celibate, in his religious idealism sacrificing his biological potential for mating and parenting (and thus his "inclusive fitness") while acting out his role as an idealized father in his heartfelt "fatherly" concern for Pompilia. In his sadistic sexuality, Guido abuses his mate and apparently is unconcerned about the birth and welfare of what the reader is asked to believe is his biological son, limiting his interest in his immediate family, ancestors, and kin relations to the genealogical heritage that allows him as an individual to claim elitist social status. In Chapter 6, I continue to discuss Browning's complex attitudes toward human nature and his representations of masculinity as I compare him with Tennyson, Barrett Browning, and Clough in more detail.

Chapter 6
Conclusion

Like other Victorian poets, Tennyson, Barrett Browning, Clough, and Browning were intensely aware of the vatic claims for poetry made by Wordsworth, Shelley, and others, and, especially in the major long poems under discussion here, they implied that they were offering their readers poetic narratives that explored universal themes of human nature. Tennyson, though he acknowledged a "parabolic drift" in *Idylls*, was careful not to imply a historical continuity between the "order" of King Arthur and that of Victorian Britain. However, he assumed that the story of Arthur and his kingdom, however exotic in its mythic setting, dealt with universal themes that explored the foundations of civilization. Barrett Browning's *Aurora Leigh* was set in the historical present, but Aurora's poetic insights into the relationships among the individual, society, and nature, and, especially, the apocalyptic vision at the conclusion of the poem, also represent engagement with universal themes. In *Amours de Voyage,* Clough acknowledges the existence of those universal themes but analyzes them from the point of view of a young man who has intellectually detached himself from vital, emotional commitments. Browning assumed that his adaptation of an Italian legal case from the late seventeenth century would be of interest to Victorians not just because of its fascinating historical setting but because it opened up issues about fundamental human values. Barrett Browning, Clough, and Browning each used an Italian setting in innovative ways that took advantage of a contemporary trend toward imaginative engagement with Italy in a time of nation-building.[1] Browning's story has a historical setting but the point of view of the contemporary, English poet–narrator is dominant. Each narrative is distanced from England and yet linked to English concerns and an English point of view in complex ways that combine the familiar with the exotic and foreign in order to invite a new vision of familiar themes. All four poets were concerned with connections between the secular and the spiritual, and all four assumed that they were doing more than participating in a Victorian culture characterized by exalted views of poetry. They were writing in a Western poetic tradition that reached back to Homer. In the words of the modern novelist Ian McEwan (who also feels connections to Homer that go beyond ideology and literary conventions), "That which binds us, our common nature, is what literature has always, knowingly and helplessly, given voice to."[2]

As argued throughout this book, the individual writer necessarily deals with impulses, drives, tendencies of human nature as he or she constructs a literary work

---

[1] See Reynolds's *The Realms of Verse, 1830–1870* for a discussion of the way Italian settings were used by these poets.

[2] See McEwan's essay "Literature, Science, and Human Nature," 19.

in a specific historical and cultural context, whether or not the writer consciously resists and rejects or idealizes aspects of human nature.

Representations of masculinity are central to the poetic narratives studied here, and I have emphasized various issues related to the construction of masculinity in the context of human nature, appealing to links between studies of human biology, the social sciences, history, and culture. It is not my intention to reduce human universals to formulas; not to discount human agency or ignore authorial moves against "natural" urges and predilections such as those associated with violence, oppression, or bigotry; not to discount human spirituality; not to ignore the reality of homosexuality among human populations that have evolved through countless generations of sexual reproduction. Instead it is to face the fact that human beings are part of the natural world. We need not theorize about the origins of art, whether it is a particular kind of adaptation within an evolutionary framework or rather a "by-product" of the evolutionary process[3] – although this kind of theorization is important and interesting – in order to acknowledge that evolutionary psychologists and cognitive scientists and anthropologists have amassed a great deal of knowledge about the "literary animal" that is relevant to discussions of art in general and literature in particular. In studying literature, an analysis of interactions between particular human cultures and human nature is always relevant, and a study of the relationship between gender and sexuality is part of that process.

In Tennyson's story about the rise and fall of Arthur's kingdom in the *Idylls*, male violence plays a central role, as I have shown in some detail. The poet began his long work with the recognition of his protagonist's fall and his consciousness of the terribly disabling, confusing, and apparently insuperable contradictions within traditions supporting the use of violence. Arthur's purpose is to employ a controlled use of violence in advancing civilization and the highest ideals, which anticipate the beautiful but finally unattainable spiritual masculinity of Christ himself, but Arthur's idealism cannot ultimately control the potential for unregulated violence within the psyches of his own knights or the instability of a social hierarchy that depends on steadfast rules governing sexual fidelity and maintenance of kinship systems subject to disruption. My assumption throughout this study has been that the regulation of a natural propensity for violence among males is one of the greatest challenges among human societies everywhere, and I have cited evidence for this generalization. Furthermore, the issue of sexual competition and jealously among males is indeed a universal one, and the most remarkable aspect of the Arthur–Guinevere–Lancelot love triangle is not that it ultimately led to the downfall of the kingdom but that Arthur and Lancelot were able to maintain a mutual respect and friendship – and shared idealism – for so long under these circumstances. A powerful irony here is that sources of great strength in Arthur – the spiritual idealism that informs his leadership and his male bonding with Lancelot that represents the height of the chivalric brotherhood associated with the Round Table – make him vulnerable to failure. As Guinevere

---

[3] See Brian Boyd's discussion of "Evolutionary Theories of Art."

remarks, to Lancelot himself, in "Lancelot and Elaine," "[W]ho loves me must have a touch of earth; / ... I am yours, / Not Arthur's, as ye know, save by the bond" (133–5), and the sexual jealousy to which Arthur is personally oblivious is a fundamental aspect of human nature: in the words of David M. Buss, "Jealousy, the dangerous passion spurred by infidelity or desertion, unleashes a fury against the partner or interloper unrivaled by any other emotion."[4] In Malory's version of the story, Arthur demands Guinevere's death by fire for treason, and Lancelot rescues her. In Tennyson's version, there is a tension in Arthur's life, not between his love for this wife and that for his friend, but between his personal commitment to his wife in their marital bond and his commitment to the social order he has founded – and ultimately his own survival as well as that of his kingdom is at issue. The childlessness of Arthur and Guinevere is also significant, as discussed in Chapter 2. The idealistic Arthur with his concept of the Order of the Round Table is in fundamental ways resisting and reforming human nature, and this helps to account for his fall.

In Barrett Browning's *Aurora Leigh,* both the protagonist and her lover are extraordinary individuals who distinguish themselves from the conventions of their society and assert themselves in ways that could be tolerated only in a relatively liberal culture that allows a good deal of individual freedom. However different in their initial positions, neither one follows the rules governing vocation or behavioral norms for individuals. Aurora refuses to become a conventional Victorian young woman, and in justifying her role as poet she defies her family and social group. Romney, in his socialist utopian philosophy and reformist projects, defies his family and social class interests as well as the dominant political ideology of his nation. But when they finally get together in the (long) ending, they not only affirm a normative heterosexual union but also as cousins in their own way reinforce continuities in kinship and social class that are consistent with universal patterns observed in cultures around the world. Significantly, Aurora recognizes in Romney a man who is willing to make a fundamental commitment to their relationship. Female mating strategies in all known cultures are largely focused on finding a mate who will make such a commitment, and Romney is an extreme case because his *investment* in Aurora is remarkable and constitutes a kind of masculine heroism. His decision to follow her poetic vision is, ironically, reinforced by his willingness to follow through in his proposal of marriage to Marian, though Marian is not interested. Barrett Browning is careful to avoid any hint of sexual jealousy in the relationship between Aurora and Marian, and the author identifies with both of them as aspects of herself. Romney is a man who can be trusted to keep his promise of commitment under any circumstances, and in spite of his unconventional ideas, he retains resources to support a potential family,

---

[4] Buss, *The Dangerous Passion*, 121. He offers an in-depth evolutionary explanation for jealousy, pointing out that it is not a "social construction." His observation that the absence of jealousy in her partner can signal lack of love to a woman (47) is relevant to Guinevere's attitude toward Arthur.

although the emphasis is on Aurora as poet–worker, as discussed in Chapter 3. The lovers' apocalyptic vision in the heavens is implicitly grounded in the real world. And love is more than a social construction.[5]

In Clough's *Amours de Voyage*, Claude becomes increasingly conscious of his role as a potential mate for Mary and ally of the Trevellyn family, and finally he desires to make this commitment, but the circumstances of life make his union with her difficult. Overriding the conventions of romantic comedy, Clough implicitly asks his readers to indulge the author in offering a problematic conclusion in which the "hero" is compromised in terms of his masculinity and the fulfillment of his romantic impulse, realistically accepts disappointment, and goes on with his life. Ironically, the problematic aspects of Claude's uncertainties and musings about his life story powerfully foreground the innate predispositions of human nature and their complex relationship with cultural formations and individual human volition as Claude reflects on questions of physical survival and potential commitments related to mating and social relations.

Browning's relentless critique of masculinity in *The Ring and the Book* grows out of an awareness of contradictions and tensions within conventional models of masculinity, as I have attempted to demonstrate in my discussion of the poem. Guido understands in a superficial way that masculinity, reinforced by an advanced position in his social hierarchy, implies certain values, including an exalted ideal of "honor" and the physical courage to defend it. Unlike Caponsacchi and the Pope, in their individual ways, however, Guido lacks the integrity and essential sense of selfhood that would enable him to at least make a sincere attempt to attain the masculine ideal which he supposedly accepts. Paradoxically, his failure to give or sacrifice himself in any way and his consistent motivation of material self-interest are products of his inner emptiness, his lack of selfhood in any meaningful sense. Browning focuses on his failure to appreciate, respect, or protect his wife Pompilia, whom he does not love.

Clearly, Browning intends to criticize the conventional (by implication the British Victorian) masculinity of his day, and, as I have argued, this critique was part of a growing consensus among Victorian *men* that women potentially suffered under the domination of men and that men should be held fully accountable in a legal and moral sense for the treatment of the women in their lives. This shift is part of a growing consciousness of individual rights that is linked to Victorian philosophical and political liberalism. In this context it is appropriate that Guido defines himself as a "natural" man. But Guido is a miserable failure in terms of all the innate predispositions of human nature related to survival, mating, parenting, kin relations, and social relations, and, of course, his general intelligence and cognitive abilities are narrowly and inadequately focused. Guido is a coward,

---

[5] Based on the findings of neuroscience, psychology, history, anthropology, philosophy, and other fields, Marcus Nordlund concludes that "love, and specifically romantic love, is not a social construction in any useful or meaningful sense of the word." See "The Problem of Romantic Love: Shakespeare and Evolutionary Psychology," 107.

and his deficiencies in terms of traditional masculinity are stressed throughout the poem, so that the poem does not (through him) condemn an archetypal, "strong" man who asserts his rights and defends his family's honor, which he believes has been challenged by his wife's infidelity. Guido himself, in Browning's version of "sexual jealousy," plots to create the illusion of his wife's faithlessness. Furthermore, Browning's construction of chivalric Christianity and his idealization of Pompilia represent his attempt to salvage connections with ancient traditions, and his valorization of *impulsive* action on the part of Caponsacchi and essential, unreflective spirituality on the part of Pompilia (who also "heroically" acts on impulse at a key point in the narrative) represent his appeal to ancient truths and a "natural" humanity underlying contemporary social conventions and reflective consciousness. Browning's associations of these phenomena with "God's truth" are probably intended to deflect the kind of criticism that George Santayana leveled at him: that his "poetry of barbarism" celebrates the "exercise of energy" as an absolute good.[6]

It is revealing to compare the four poems in terms of the concept of masculine "chivalry." As shown in Chapter 2, chivalry is an important key to regulating the male behavior that supports Arthur's "order," managing masculine violence and maintaining a civilized code of honor. At the same time, however, it can encourage adultery, and in the final analysis Tennyson demonstrates that the protection, rescue, and idealization of women prove an inadequate disciplinary principle. Browning's version of chivalrous Christianity, however, is validated in his poem, and the idealization of Pompilia is a key to the values implicit in *The Ring and the Book*. In contrast, Barrett Browning rejects the principle of chivalry by refusing to let Romney, however dedicated and willing to sacrifice himself, rescue either Marian or Aurora according to the usual conventions of chivalry. And although Aurora herself insists at the end that love rather than poetry furnishes the key to highest human value, Romney's idealization of her is focused on her poetry as well as her person. I have argued that in some ways Romney is a knight-like hero when he comes to Aurora, but Barrett Browning takes care to avoid valorizing the use of physical force by men. At the same time, however, she does not emphasize male violence in a negative context nearly as much as Tennyson and Browning do. For them, male violence drives the plots of their narratives. Barrett Browning includes the rape of Marian as a key element in the plot but does not dramatize or even describe it very clearly as an incident. Marian's father is a violent man, but neither his abuse of Marian nor his part in the blinding of Romney is described in any detail either. While Tennyson and Browning focus on violent male–male competition, Barrett Browning focuses on the nonviolent but psychologically intense female–female competition between Aurora and Lady Waldemar. In Clough's poem, Claude declines commitment to a meaningful role as warrior or lover though both the setting of the work and its poetic form suggest traditions of masculine heroism that set off his intellectual passivity and Claude

---

[6] In his essay "The Poetry of Barbarism" (1900).

himself acknowledges the pull of human nature and finally asserts himself in a desperate attempt to overtake the Trevellyn family and claim Mary as his mate before acknowledging failure and accepting his status as a kind of intellectual anti-hero, with a tendency toward cowardice and ill-defined links to family and nation.

The exemplary masculinity of a principal character is central to the plot and to any comprehensive interpretation of the narrative poems by Tennyson, Barrett Browning, and Browning, but in each case such masculinity is complex, and the intricacies involved in defining and assessing it are closely related to traditional literary concerns. Tennyson's King Arthur, representing "[i]deal manhood closed in real man," struggles to develop and lead an order that incorporates what he takes to be the highest spiritual ideals; but his order collapses around him, not because it is defeated by "heathen" enemies but because of internal fractures within his own family and among his closest friends and allies. The hierarchical structure of the Round Table cannot survive when the "cycle" of Arthur's kingdom has run its course, and there is no heir to succeed Arthur. The ideology of chivalry as defined by Arthur was an invaluable tool in reinforcing and expanding his order, but it was not adequate to hold it in place when his own familial and tribal structures disintegrated. Tennyson meant to represent Arthur as a truly exceptional man, but, in spite of the (inherited) associations with pagan and Christian mythology, this character is most interesting as a human being who struggles to fulfill a mission he does not fully understand. Clough's somewhat comic questioning of nationalism and chivalric idealism offers a striking contrast to Tennyson in terms of point of view and literary genre but in its own way powerfully questions conventional concepts of masculinity.

Neither Barrett Browning's Romney Leigh nor Browning's Giuseppe Caponsacchi has the stature of Tennyson's King Arthur, but each character is of vital importance in the context of the poem in which he appears. Far from offering an "opposition between triumphant female present and desiccated male past"[7] in *Aurora Leigh*, Barrett Browning celebrates the universal, biological, and spiritual love between men and women that transcends history. In the poet's scheme of things, Romney's masculinity has indeed been misguided in his championing absurd ideological schemes and in his failure to understand the centrality of *poetry* as a way to spiritual truth, but his heroic return to Aurora, his commitment to her future with all that implies, and his fulfillment of the prophecy of Aurora's father confirm his identity as Aurora's ideal mate. And Browning's Caponsacchi, in stark contrast to his monstrous rival Guido, consciously reconstructs his life, adopting a kind of chivalrous Christianity that allows him to find a meaningful masculine role while retaining his status as a celibate priest. Both Caponsacchi and Browning understand that this formulation is tenuous indeed. To accommodate the women they love, Romney Leigh and Giuseppe Caponsacchi radically reformulate their identities as individuals within male status hierarchies. It is critical to Browning's

---

[7] See Auerbach, 164.

scheme that Caponsacchi retain his celibacy because Guido's appeal to his *natural* right to experience jealousy and seek revenge must clearly be marked as a *lie* in order to preserve the poet's critical point of view. Guido himself flouts the universals of human nature as conceptualized by Browning. Guido is of course "unnatural" and more "unmanly" than Caponsacchi, and, ironically, it is the celibate Pope who communicates a universal, patriarchal view of the world that, somewhat like that of Arthur, recognizes the value of physical courage in males, especially in defending innocent women, though it condemns the unjustifiable male violence that infects the world.

All four poets assumed that complex concepts of masculinity are inextricably connected with the human themes they wished to develop in their epic narratives and that biological and cultural issues are interrelated. In order to study their literary works in a meaningful way, we should search for methods of linking what we know about Victorian literature, culture, and society with the universals of human nature as defined by scientific studies, especially those based on adaptationist psychology.

Bibliography

References to Tennyson's poetry given in the text are based on *The Poems of Tennyson*, edited by Christopher Ricks, 3 vols (Berkeley: University of California Press, 1987). References to Barrett Browning's *Aurora Leigh* are based on *Aurora Leigh*, edited by Margaret Reynolds (Athens: Ohio University Press, 1992) and references to other works by her are based on *The Complete Works of Elizabeth Barrett Browning*, edited by Charlotte Porter and Helen A Clarke (New York: Thomas Y. Cromwell, 1900; rpt New York: AMS Press, 1973). References to Clough's *Amours de Voyage* are based on *Amours de Voyage*, edited by Patrick Scott (St. Lucia: University of Queensland Press, 1974), and references to other works by him are based on *The Poems of Arthur Hugh Clough*, second edition, edited by F. L. Mulhauser (Oxford: Clarendon Press, 1974). References to Browning's poetry are based on *The Complete Works of Robert Browning: With Variant Readings and Annotations,* ed. Roma A. King, Jr. et al., 13 vols (Athens: Ohio University Press, 1969–1998).

Adams, James Eli. *Dandies and Desert Saints: Styles of Victorian Masculinities.* Ithaca: Cornell University Press, 1995.

Adams, Michael C. C. *The Great Adventure: Male Desire and the Coming of World War I.* Bloomington: Indiana University Press, 1990.

Alaya, Flavia."The Ring, the Rescue, & the Risorgimento Reunifying the Brownings' Italy," *Browning Institute Studies* 6 (1978): 17–20.

Altick, Richard D. "The Private Life of Robert Browning." *Yale Review* 41 (1951): 247–62.

Altick, Richard D., and James F. Loucks, II. *Browning's Roman Murder Story.* Chicago: University of Chicago Press, 1968.

Appleton, Jay. *The Symbolism of Habitat: An Interpretation of Landscape in the Arts.* Seattle: University of Washington Press, 1990.

Armstrong, Isobel. *Arthur Hugh Clough.* London: Longmans, Green, 1962.

———. *Victorian Poetry: Poetry, Poetics and Politics.* London: Routledge, 1993.

Arnold, Matthew. *The Complete Prose Works of Matthew Arnold.* Ed. R. H. Super. 11 vols. Ann Arbor: University of Michigan Press, 1960–1977.

———. *The Letters of Matthew Arnold to Arthur Hugh Clough.* Ed. Howard Foster Lowry. London: Oxford University Press, 1932.

Auerbach, Nina. "Robert Browning's Last Word." *Victorian Poetry* 22 (1984): 161–73.

Avery, Simon, and Rebecca Stott. *Elizabeth Barrett Browning.* London: Pearson, 2003.

Aytoun, William Edmonstoune. *Firmilian, or the Student of Badajoz: A Spasmodic Tragedy.* London: William Blackwood and Sons, 1854.

———. "Mrs. Barrett's *Aurora Leigh*." *Blackwood's Edingurgh Magazine*. January 1857: 23–41.
Barash, David P., and Nanelle R. Barash. *Madame Bovary's Ovaries: A Darwinian Look at Literature*. New York: Delacorte, 2005.
Barrett Browning, Elizabeth. *The Letters of Elizabeth Barrett Browning*. Ed. Frederick G. Kenyon. 2 vols. London: Macmillan, 1897.
Beer, Gillian. *Darwin's Plots: Evolutionary Narrative in Darwin, George Eliot and Nineteenth-Century Fiction*. London: Routledge and Kegan Paul, 1985.
Besant, Walter. *The Revolt of Man*. Edinburgh: Blackwood, 1882.
Biswas, Robindra Kumar. *Arthur Hugh Clough: Towards a Reconsideration*. Oxford: Clarendon Press, 1972.
Blair, Kirstie. "Spasmodic Affections: Poetry, Pathology, and the Spasmodic Hero." *Victorian Poetry* 42 (2004): 473–90.
Boos, Florence S. "'Spasm' and Class: W. E. Aytoun, George Gilfillan, Sydney Dobell, and Alexander Smith." *Victorian Poetry* 42 (2004): 553–83.
Booth, Wayne C. *The Rhetoric of Fiction*. 2nd edition, 1983. University of Chicago Press, 1961.
Boyd, Brian. "Evolutionary Theories of Art." In *The Literary Animal: Evolution and the Nature of Narrative*. Eds. Jonathan Gottschall and David Sloan Wilson. Evanston, IL: Northwestern University Press, 2005. 147–76.
———. "Getting It All Wrong," *The American Scholar* 75.4 (2006): 18–30.
———. "Literature and Evolution: A Bio-Cultural Approach," *Philosophy and Literature* 29 (2005): 1–23.
———. *On the Origin of Stories: Evolution, Cognition and Fiction*. Cambridge, MA: Harvard University Press, 2009.
———. "The Origin of Stories: *Horton Hears a Who*." *Philosophy and Literature* 25.2 (2001): 197–214.
Brady, Ann P. *"Pompilia": A Feminist Reading of Robert Browning's "The Ring and the Book."* Ohio University Press, 1988.
Brisbane, Thomas. *The Early Years of Alexander Smith, Poet and Essayist*. London: Hodder & Stoughton, 1869.
Brooks, Peter. *The Melodramatic Imagination*. New Haven, CT: Yale University Press, 1976.
Brown, Donald E. *Human Universals*. New York: McGraw Hill, 1991.
———. "Human Universals and Their Implications." In *Being Humans: Anthropological Universality and Particularity in Transdiscilinary Perspectives*. New York: Walter de Gruyter, 2000.
Browning, Robert, and Elizabeth Barrett Browning. *The Brownings' Correspondence*. vol. 9. Eds. Philip Kelly and Ronald Hudson. Winfield, KS: Wedgestone Press, 1991.
———. *The Letters of Robert Browning and Elizabeth Barrett Browning 1845–1846*. Ed. Elvan Kintner. Cambridge, MA: Harvard University Press, 1969.
Buckler, William E. *Poetry and Truth in Robert Browning's "The Ring and the Book."* New York: New York University Press, 1985.

Buckley, Jerome H. *The Victorian Temper, A Study in Literary Culture*. New York: Vintage, 1951.

Buller, David J. *Evolutionary Psychology and the Persistent Quest for Human Nature*. Cambridge, MA: MIT Press, 2005.

Buss, David. M. *The Evolution of Desire: Strategies in Human Mating*. New York: Basic Books, 1994.

———. *The Dangerous Passion: Why Jealousy Is as Necessary as Love and Sex*. New York: Free Press, 2000.

Carpenter, Mary Wilson. "Blinding the Hero." *Differences* 17.3 (2006): 52–68.

Carroll, Joseph. *Evolution and Literary Theory*. Columbia: University of Missouri Press, 1995.

———. "An Evolutionary Paradigm for Literary Study." *Style* 42.2&3 (2008): 103–41.

———. "Human Nature and Literary Meaning: A Theoretical Model Illustrated with a Critique of *Pride and Prejudice*." In *The Literary Animal: Evolution and the Nature of Narrative*. Eds. Jonathan Gottschall and David Sloan Wilson. Evanston, IL: Northwestern University Press, 2005. 76–106.

———. "The Human Revolution and the Adaptive Function of Literature." *Philosophy and Literature* 30 (2006): 33–49.

———. *Literary Darwinisn: Evolution, Human Nature, and Literature*. New York: Routledge, 2004.

Case, Alison. "Gender and Narration in *Aurora Leigh*," *Victorian Poetry* 29 (1): 25–32.

Christ, Carol. "Victorian Masculinity and the Angel in the House." In *A Widening Sphere*. Ed. Martha Vincinus. Bloomington: Indiana University Press, 1977. 146–62.

Chorley, Katharine. *Arthur Hugh Clough: The Uncommitted Mind, A Study of His Life and Poetry*. Oxford: Clarendon Press, 1962.

———. *Victorian and Modern Poetics*. 1984. Chicago: University of Chicago Press, 1984.

Christiansen, Rupert. *The Voice of Victorian Sex: Arthur Hugh Clough*. London: Short Books, 2001.

Clough, Arthur Hugh. *The Correspondence of Arthur Hugh Clough*. 2 vols. Ed. F. L. Mulhauser. Oxford: Clarendon Press, 1957.

———. *The Poems and Prose Remains of Arthur Hugh Clough*, Ed. by his wife. 2 vols. London: Macmillan, 1869.

Conrad, Joseph. *Heart of Darkness*. London: Penguin, 1995. Originally published 1899.

Cook, A. K. *A Commentary upon Browning's 'The Ring and the Book'*. London: Oxford University Press, 1920.

Cooke, Brett, ed. "Literary Biopetics." Special Issue. *Interdisciplinary Literary Studies* 2.2 (2001).

Cooke, Brett, and Frederick Turner, eds. *Evolutionary Explorations in the Arts*. Lexington, KY: International Conference on the Unity of the Sciences, 1999.

Cooper, Hellen. *Elizabeth Barrett Browning, Woman and Artist*. University of North Carolina Press, 1988.

Corrigan, Beatrice. *Curious Annals: New Documents Relating to Browning's Roman Murder Story*. Toronto: University of Toronto Press, 1956.

Cott, Nancy F. "Passionlessness: An Interpretation of Victorian Sexual Ideology, 1790–1850." *Signs* 4 (1978): 219–36.

Crowder, Ashby. "Browning and Women," *Studies in Browning and his Circle* 14 (1986): 91–134.

Dalley, Lana L. "The least 'Angelical' poem in the language": Political Economy, Gender, and the Heritage of *Aurora Leigh*." *Victorian Poetry* 44.4 (2006): 525–45.

Daly, M., and M. Wilson. *Homicide*. Hawthorne, NY: Aldine deGruyter, 1988.

David, Deirdre. *Intellectual Women and Victorian Patriarchy: Harriet Martineau, Elizabeth Barrett Browning, George Eliot*. London; Macmillan, 1987.

Delmas, Phillipe. *The Rosy Future of War*. New York: Free Press, 1997.

Dennet, William C. *Freedom Evolves*. New York: Viking, 2003.

Dennis, Barbara. *Elizabeth Barrett Browning: The Hope End Years*. Bridgend: Seren, 1996.

Dentith, Simon. *Epic and Empire in Nineteenth-Century Britain*. Cambridge: Cambridge University Press, 2006.

DeVane, William C. "The Virgin and the Dragon." *Yale Review* 37 (1947): 33–46.

Digby, Kenelm Henry. 1845–76. *The Broad Stone of Honour; or, The True Sense and Practice of Chivalry*. 4 vols. London: Edward Lumley, 1845–76.

Dissanayake, Ellen. *Art and Intimacy: How the Arts Began*. Seattle: University of Washington Press, 2000.

———. *Homo Aestheticus: Where Art Comes From and Why*. Seattle: University of Washington Press, 1995.

Dunbar, Robin. "The Social Brain Hypothesis." *Evolutionary Anthropology* 6 (1998): 178–90.

Duncan, David. *The Life and Letters of Herbert Spencer*. London: Methuen, 1908.

Dutton, Denis. *The Art Instinct: Beauty, Pleasure, and Human Evolution*. New York: Bloomsbury Press, 2009.

Easterlin, Nancy. "Hans Christian Andersen's Fish Out of Water." *Philosophy and Literature* 25.2 (2001): 251–77.

———. "How to Write the Great Darwinian Novel: Cognitive Presidspositions, Cultural Complexity, and Aesthetic Evaluation." *The Journal of Evolutionary and Cultural Psychology* 3.1 (2005): 23–8.

Ehrenreich, Barbara. *Blood Rites: Origins and History of the Passions of War*. New York: Metropolitan, 1997.

Eliot, T. S. "*In Memoriam*." In *Essays Ancient and Modern*. London: Faber and Faber, 1936. 175–90.

Farrell, John P. "'The Scholar-Gipsy' and the Continuous Life of Victorian Poetry." *Victorian Poetry* 43 (2005): 277–96.

Fauconnier, Gilles, and Mark Turner. *The Way We Think: Conceptual Blending and the Mind's Hidden Complexities*. New York: Basic Books, 2002.

Foucault, Michel. *The History of Sexuality: Vol. 1, an Introduction*. Harmondsworth: Penguin, 1984.

Fox, Robin. "Male Boonding in the Epics and Romances." In *The Literary Animal: Evolution and the Nature of Narrative*. Eds. Jonathan Gottschall and David Sloan Wilson. Evanston, IL: Northwestern University Press, 2005. 126–44.

Freud, Sigmund. "The Antithetical Sense of Primal Words." In *Character and Culture*. Ed. Philip Rieff. New York: Collier, 1963.

Fussell, Paul. *The Great War and Modern Memory*. London: Oxford University Press, 1975.

Gavins, Joanna, and Gerald Steen, eds. *Cognitive Poetics in Practice*. London: Routledge, 2003.

Geary, David C. *Male, Female: The Evolution of Sex Differences*. Washington, DC: American Psychological Association, 1998.

———. *The Origin of Mind: Evolution of Brain, Cognition, and General Intelligence*. Washington, DC: American Psychological Association, 2005.

———. "Sexual Selection and Human Life History." In *Advances in Child Development and Behavior*, vol. 30, 41–101. Ed. R. Kail. San Diego, Academic Press, 2002.

———. "Sexual Selection and Sex Differences in Social Cognition." In *Biology, Sociology, and Behavior: The Development of Sex Differences in Cognition*, 23–53. Eds. A. V. McGillicuddy-DeLisi and R. De-Lisi. Greenwich: Ablex/Greenwood, 2002.

Gest, John Marshall, ed. and trans. *The Old Yellow Book: Source of Browning's "The Ring and The Book."* Boston: Chipman Law Publishing Company, 1925.

Ghiglicri, M. P. "Sociobiology of the Great Apes and the Hominid Ancestor" *Journal of Human Evolution* 16 (1987): 319–57.

Gibson, Mary Ellis, ed.. *Critical Essays on Robert Browning*. New York: G. K. Hall, 1992.

Gilbert, Elliot L. "The Female King: Tennyson's Arthurian Apocalypse." *PMLA* 98 (1983): 863–78.

Girouard, Mark. *The Return to Camelot*. New Haven, CT: Yale University Press, 1981.

Gottschall, Jonathan. "Homer's Human Animal: Ritual Combat in the *Iliad*. *Philosophy and Literature* 25.2 (2001): 278–94.

———. *Literature, Science, and a New Humanities*. New York: Palgrave Macmillan, 2008.

———. *The Rape of Troy: Evolution, Violence and the World of Homer*. Cambridge: Cambridge University Press, 2008.

———. "The Tree of Knowledge and Darwinian Literary Studies." *Philosophy and Literature* 27 (2003): 255–268.

Gottschall, Jonathan and David Sloan Wilson, eds. *The Literary Animal: Evolution and the Nature of Narrative*. Evanston, IL: Northwestern University Press, 2005.

Graham, Colin. *Ideologies of Epic: Nation, Empire and Victorian Epic Poetry.* Manchester: Manchester University Press, 1998.

Gray, Chris Hables. *Postmodern War: The New Politics of Conflict.* New York: Guilford Press, 1997.

Gray, J. M. 1980. *Thro' the vision of the night: A Study of Source, Evolution, and Structure in Tennyson's "Idylls of the King."* Montreal: McGill-Queen's University Press, 1980.

Greenberger, Evelyn Barish. *Arthur Hugh Clough: The Growth of a Poet's Mind.* Cambridge, MA: Harvard University Press, 1970.

Hanley, Evelyn A. *The Subjective Vision: Six Victorian Women Poets.* New York: Astra, 1978.

Harris, Wendell V. *Arthur Hugh Clough.* New York: Twayne Publishers, 1970.

Harrison, Antony H., and Beverly Taylor, eds. *Gender and Discourse in Victorian Literature and Art.* DeKalb: Northern Illinois University Press, 1992.

Harrison, Antony H. "Victorian Culture Wars: Alexander Smith, Arthur Hugh Clough, and Matthew Arnold in 1853." *Victorian Poetry* 42 (2004): 509–20.

Hogan, Patrick Colm. *The Mind and Its Stories: Narrative Universals and Human Emotion.* Cambridge: Cambridge University Press, 2003.

Holloway, John. *The Victorian Sage, Studies in Argument.* London: Macmillan, 1953.

Honan, Park. *Browning's Characters: A Study in Poetic Technique.* New Haven, CT: Yale University Press, 1961.

Houghton, Walter E. *The Poetry of Clough: An Essay in Revaluation.* New Haven, CT: Yale University Press, 1963.

———. *The Victorian Frame of Mind, 1830-1870.* New Haven, CT: Yale Univeristy Press, 1957.

Hughes, Linda K. "Alexander Smith and the Bisexual Poetics of *A Life-Drama.*" *Victorian Poetry* 42 (2004): 491–508.

———. "Introduction." *Victorian Poetry* 41.4 (2003): 459–63.

———. "Whithering Away: Editorial Introduction." *Victorian Poetry* 42.1 (2004): 1–8.

Ignatieff, Michael. "The Gods of War." *The New York Review of Books* (9 October 1997): 10–13.

Irvine, William, and Park Honan. *The Book, the Ring, and the Poet.* New York: McGraw-Hill, 1974.

James, Henry. "Tennyson's Drama." *Views and Reviews.* Boston: Ball, 1908.

Johnson, E. D. H. *The Alien Vision of Victorian Poetry: Sources of the Poetic Imagination in Tennyson, Browning, and Arnold.* Princeton, NJ: Princeton University Press, 1952.

———. Robert Browning's Pluralistic Universe: A Reading of *The Ring and the Book. University of Toronto Quarterly* 31 (1961): 20–21.

Johnson, Stephanie L. "*Aurora Leigh*'s Radical Youth: Derridean *Parergon* and the Narrative Frame in "A Vision of Poets." *Victorian Poetry* 44.4 (2006): 425–44.

Jones, Steven, Robert Martin, and David Philbeam. *The Cambridge Encyclopedia of Human Evolution*. Cambridge: Cambridge University Press, 1992.

Joseph, Gerhard. *Tennysonian Love: The Strange Diagonal*. Minneapolis: University of Minnesota Press, 1969.

Kaplan, Cora. "Introduction" to Elizabeth Barrett Browning, *"Aurora Leigh" With Other Poems*. London: Women's Press, 1978.

Kaplan, Hillard, Kim Hill, Jane Lancaster, and A. Magdalena Hurtado. 2000. "A Theory of Human Life History Evolution: Diet, Intelligence, and Longevity." *Evolutionary Anthropology* 9 (2005): 156–85.

Kenny, Anthony. *Arthur Hugh Clough: A Poet's Life*. London: Continuum, 2005.

Kiernan, V. G. *The Duel in European Hisotry: Honour and the Reign of Aristocracy*. Oxford; Oxford University Press, 1967.

Kingsley, Charles. "Thoughts on Shelley and Byron." *Fraser's Magazine* 49 (1853): 568–576.

Keeley, L. H. *War Before Civilization: The Myth of the Peaceful Savage*. New York: Oxford University Press, 1996.

Klein, Richard G. "Archeology and the Evolution of Human Behavior." *Evolutionary Anthrobiology* 9 (2000): 17–36.

Knoepflmacher, U. C. "Idling in Gardens of the Queen: Tennyson's Boys, Princes, and Kings." *Victorian Poetry* 30 (1992): 343–64.

———. "Projection and the Female Other: Romanticism, Browning, and the Victorian Dramatic Monologue." In *Critical Essays on Robert Browning*. Ed. Mary Ellis Gibson. New York: G. K. Hall, 1992. 100–19.

Kruger, Daniel J., Maryanne Fisher, and Ian Jobling. "Proper Hero Dads and Dark Hero Cads: Alternate Mating Strategies Exemplified in British Romantic Literature." In *The Literary Animal: Evolution and the Nature of Narrative*. Eds. Jonathan Gottschall and David Sloan Wilson. Evanston, IL: Northwestern University Press, 2005. 225–43.

Lakoff, George, and Mark Johnson. *Metaphors We Live By*. Chicago: University of Chicago Press, 1980.

Langbaum, Robert. *The Poetry of Experience: The Dramatic Monologue in Modern Literary Tradition*. New York: Random House, 1957.

———. "Browning and the Question of Myth. *PMLA* 81.7 (1966): 575–84.

LaPorte, Charles, "Spasmodic Poetics and Clough's Apostacies." *Victorian Poetry* 42 (2004): 521–36.

Leighton, Angela. 1986. *Elizabeth Barrett Browning*. Brighton: Harvester Press, 1986.

Levine, George, and William Madden, eds. *The Art of Victorian Prose*. New York: Oxford University Press, 1968.

Love, Glen A. *Practical Ecocriticism: Literature, Biology, and the Environment*. Charlottesville: University of Virginia Press, 2003.

Machann, Clinton. "The Male Villain as Domestic Tyrant in *Daniel Deronda*: Victorian Masculinities and the Cultural Context of George Eliot's Novel." *Journal of Men's Studies* 13:3 (2005), 327–46.

———. *Matthew Arnold: A Literary Life*. New York: St. Martin's, 1998.

Mason, Emma. "Rhythmic Numinousness: Sydney Dobell and 'The Church.'" *Victorian Poetry* 42 (2004): 537–51.

McEwan, Ian. "Literature, Science, and Human Nature." In *The Literary Animal: Evolution and the Nature of Narrative*. Eds. Jonathan Gottschall and David Sloan Wilson. Evanston, IL: Northwestern University Press, 2005. 5–19.

McGuire, Ian. "Epistemology and Empire in *Idylls of the King*." *Victorian Poetry* 30 (1992): 387–400.

McNally, James. "Touches of Aurora Leigh in *The Ring and the Book*," *Studies in Browning and His Circle* 14 (1986): 85–90.

McSweeny, Kerry. *Tennyson and Swinburne as Romantic Naturalists*. University of Toronto Press, 1981.

Mermin, Dorothy. *Elizabeth Barrett Browning: The Origins of a New Poetry*. Chicago: University of Chicago Press, 1989.

Miller, Betty. *Robert Browning: A Portrait*. New York: Scribners, 1952.

Morgan, Thaïs E. "The Poetry of Victorian Masculinities." In *The Cambridge Companion to Victorian Poetry*. Ed. Joseph Bristow. Cambridge: Cambridge University Press, 2000. 203–27.

Munich, Adrienne. *Andromeda's Chains: Gender and Interpretation in Victorian Literature and Art*. New York: Columbia University Press, 1989.

Nelson, Claudia. *Invisible Men: Fatherhood in Victorian Periodicals, 1850–1910*. Athens: University of Georgia Press, 1995.

———. "Deconstructing the Paterfamilias: British Magazines and the Imagining of the Maternal Father, 1850–1910." *Journal of Men's Studies* 11 (2004): 293–308.

Nisbett, R. E., and D. Cohen. *Culture of Honour: The Psychology of Violence in the South*. New York: HarperCollins, 1996.

Nordlund, Marcus. "Consilient Literary Interpretation." *Philosophy and* Literature 26 (2002): 312–33.

———. "The Problem of Romantic Love: Shakespeare and Evolutionary Psychology." In *The Literary Animal: Evolution and the Nature of Narrative*. Eds. Jonathan Gottschall and David Sloan Wilson. Evanston, IL: Northwestern University Press, 2005. 107–25.

O'Neill, P. *Robert Browning and Twentieth-Century Criticism*. Columbia, SC: Camden House, 1995.

Paden, W. D. *Tennyson in Egypt: A Study of the Imagery in His Earlier Work*. New York: Octagon, 1971.

Pater, Walter. *The Renaissance: Studies in Art and Poetry*. New York: Oxford University Press, 1986.

Phillips, Catherine. 2002. "'Charades from the Middle Ages?' Tennyson's *Idylls of the King* and the Chivalric Code." *Victorian Poetry* 40 (2002): 241–53.

Pinker, Steven. *The Blank Slate: The Modern Denial of Human Nature*. New York: Viking Penguin, 2002.

Plavcan, J. M., C. P. Schaik, and P. M. Kappler, J. M., C. P. Schaik, and P. M. Kappler. "Competition, Coalitions and Canine Size in Primates." *Journal of Human Evolution* 28 (1995): 245–76.

Potts, Richard. "Variability Selection in Hominid Evolution." *Evolutionary Anthropology* 7 (1998): 81–95.
Priestley, F. E. L. "Tennyson's *Idylls*." In *Critical Essays on the Poetry of Tennyson*. Ed. John Killham. New York: Barnes & Noble, 1960. 239–55.
Rampton, David. "Back to the Future: Lionel Trilling, 'The Scholar-Gipsy,' and the State of Victorian Poetry." *Victorian Poetry* 45.1 (2007):1–15.
Reynolds, Matthew. *The Realms of Verse 1830–1870: English Poetry in a Time of Nation Building*. Oxfrod: Oxford University Press, 2001.
Rhoads, Steven E. *Taking Sex Differences Seriously*. San Francisco: Encounter, 2004.
Ricks, Christopher. *Tennyson*. Berkeley: University of California Press, 1989.
Ridenour, George M. Robert Browning and *Aurora Leigh*," *Victorian Newsletter* 67 (1985): 26–32.
Roberts, Adam. *Romantic and Victorian Long Poems: A Guide*. Aldershot: Ashgate Publishing, 1999.
Robinson, Henry Crabb. *Henry Crabb Robinson on Books and their Writers*. Ed. E. J. Morley. 2 vols. London: J. M. Dent, 1938.
Rosenberg, John D. *The Fall of Camelot: A Study of Tennyson's "Idylls of the King"*. Cambridge, MA: Belknap-Harvard University Press, 1973.
Rudy, Jason R. "Rhythmic Intimacy, Spasmodic Epistemology." *Victorian Poetry* 42 (2004): 451–72.
Ruskin, John. *The Works of John Ruskin*. Eds. E. T. Cook and Alexander Wedderburn. 39 vols. London: George Allen, 1909.
Ryals, Clyde de L. *From the Great Deep: Essays on Idylls of the King*. Athens: Ohio University Press, 1967.
Ryan, Vanessa. "Why Clough? Why Now? (Arthur Hugh Clough)." *Victorian Poetry* 41.1 (2003): 504–9.
Santayana, George. "The Poetry of Barbarism." In *Interpretations of Poetry and Religion*. New York: Scribner, 1900. 188–216.
Schatz, Sueann. 2000. "*Aurora Leigh* as Paradigm of Domestic-Professional Fiction." *Philological Quarterly* 79.1 (2000): 91–117.
Sedgwick, Eve Kosofsky. *Between Men: English Literature and Male Homosocial Desire*. New York: Columbia University Press, 1985.
Shaw, Marion. *Alfred Lord Tennyson*. Atlantic Highlands, NJ: Humanities Press International, 1988.
Shaw, W. David. *The Lucid Veil: Poetic Truth in the Victorian Age*. London: Athlone Press, 1987.
———. "Browning's Murder Mystery: *The Ring and the Book* and Modern Theory." *Victorian Poetry* 27 (1989): 79–98.
Shires, Linda. "Patriarchy, Dead Men, and Tennyson's *Idylls of the King*. *Victorian Poetry* 30 (1992): 401–19.
Simpson, Roger. *Camelot Regained: The Arthurian Revival and Tennyson, 1800–1849*. Cambridge: D. S. Brewer, 1990.
Sinfield, Alan. *Alfred Tennyson*. Oxford: Basil Blackwell, 1986.

Slingerland, Edward. *What Science Offers the Humanities: Integrating Body and Culture*. Cambridge: Cambridge University Press, 2008.

Slinn, E. Warwick. *Browning and the Fictions of Identity*. Totowa, NJ: Barnes and Noble, 1982.

———. *The Discourse of Self in Victorian Poetry*. Charlottesville: University of Virginia Press, 1991.

———. 1989. "Language and Truth in *The Ring and the Book*." *Victorian Poetry* 27.3&4 (1989):115–33.

Smith, Alexander. *A Life-Drama and Other Poems*. Boston: Ticknor and Fields, 1859.

Smuts, B. B., D. L. Cheney, R. M. Seyfarth, R. W. Wrangham, and T. T. Struhsaker, eds. *Primate Societies*. Chicago: University of Chicago Press, 1987.

Stockwell, Peter. 2002. *Cognitive Poetics: An Introduction*. London: Routledge, 2002.

Stone, Marjorie. *Elizabeth Barrett Browning*. New York: St. Martin's Press, 1995.

Sussman, Herbert. *Victorian Masculinities*. Cambridge: Cambridge University Press, 1995.

Sugiyama, Michelle Scalise. "Narrative Theory and Function: Why Evolution Matters." *Philosophy and Literature* 25.2 (2001): 233–50.

Swinburne, Algernon Charles. "A. C. Swinburne on the *Idylls*." In *Tennyson: The Critical Heritage*. Ed. John D. Jump. London: Routledge, 1967. 318–21.

———. "A. C. Swinburne Replies to Taine." In *Tennyson: The Critical Heritage*. Ed. John D. Jump. London: Routledge, 1967. 336–47.

Tanner, J. M. "Human Growth and Development." In *The Cambridge Encyclopedia of Human Evolution*. Eds. S. Jones, R. Martin, and D. Philbeam. New York: Cambridge University Press, 1992. 98–105.

Taplin, Gardner B. 1957. *The Life of Elizabeth Barrett Browning*. New Haven, CT: Yale University Press, 1957.

Tennyson, Hallam. *Alfred Lord Tennyson: A Memoir by His Son*. 2 vols. London, 1897.

Thorpe, Michael, ed. *Clough: The Critical Heritage*. New York: Barnes & Noble, 1972.

Tiger, Lionel. *Men in Groups*. London: Thomas Nelson, 1968.

Timko, Michael. *Innocent Victorian: The Satiric Poetry of Arthur Hugh Clough*. Columbus: Ohio University Press, 1963.

Tosh, John. 1999. *A Man's Place: Masculinity and the Middle-Class Home in Victorian England*. New Haven, CT: Yale University Press, 1999.

Tucker, Herbert F. *Browning's Beginnings: The Art of Disclosure*. Minneapolis: University of Minnesota Press, 1980.

———. *Epic: Britain's Heroic Muse 1790–1910*. Oxford: Oxford University Press, 2008.

———. "Glandular Omnism and Beyond: The Victorian Spasmodic Epic." *Victorian Poetry* 42 (2004): 429–50.

Wade, Nicholas. *Before the Dawn: Recovering the Lost History of Our Ancestors*. New York: Penguin, 2006.

Weeks, Jeffrey. *Sex, Politics and Society: The Regulation of Sexuality since 1800*. London: Longmans, 1981.

Weinstein, Mark A. *William Edmondstoune Aytoun and the Spasmodic Controversy*. New Haven, CT: Yale University Press, 1968.

Wiener, Martin J. "The Victorian Criminalization of Men." In *Men and Violence: Masculinity, Honor Codes and Violent Rituals in Europe and America, 1600–2000*. Ed. P. Spierenburg. Columbus: Ohio State University Press, 1997.

———. *Men of Blood: Violence, Manliness and Criminal Justice in Victorian England*. Cambridge: Cambridge University Press, 2004.

Wilson, E. O. *Consilience: The Unity of Knowledge*. New York: Knopf, 1998.

———. "Foreword." In *The Literary Animal: Evolution and the Nature of Narrative*. Eds. Jonathan Gottschall and David Sloan Wilson. Evanston, IL: Northwestern University Press, 2005. vii–xi.

Woolford, John, and Daniel Karlin. *Robert Browning*. London: Longman, 1996.

Wörn, Alexandra M. B. "'Poetry is where God is': The Importance of Christian Faith and Theology in Elizabeth Barrett Browning's Life and Work." In *Victorian Religious Discourse: New Directions in Elizabeth Barrett Browning's Life and Work*. Ed. Jude V. Nixon. New York: Palgrave Macmillan, 2004. 253–72.

Zonana, Joyce. "The Embodied Muse: Elizabeth Barrett Browning's *Aurora Leigh* and Feminist Poetics." *Tulsa Studies in Women's Literature* 8 (1989): 241–62.

Zunshine, Lisa. "Eighteenth-Century Print Culture and the 'Truth' of Fictional Narrative." *Philosophy and Literature* 25.2 (2001): 215–32.

———. *Why We Read Fiction: Theory of Mind and the Novel*. Columbus: Ohio State University Press, 2006.

Index

Adams, James Eli 14, 15n, 16, 38*n*, 42, 48*n*
Adams, Michael C. C. 51*n*
adaptationist psychology 16, 21, 22, 23, 24, 27, 142, 147; *see also* evolutionary psychology; cognitive science
Alaya, Flavia 120*n*
Altick, Richard 112, 119, 133n
Ambarvalia (Clough) 83, 100
Amours de Voyage (Clough) 1, 4, 9, 11, 12, 13, 26, 83–108, 114, 139, 141, 144; *see also* chivalry; cowardice; heroism; male violence
    controversial ending of 92, 93, 94, 95, 96, 97, 98, 99, 101, 102, 107
    English national identity and familial loyalty in 84, 85, 86, 87, 88, 90, 92, 96, 97, 102, 106, 107
    flawed and conflicted masculinity in 86, 87, 89, 90, 97, 106, 107
    frustrated courtship and marriage plot in 84, 86, 88, 89, 90, 92, 93, 94, 95, 96, 97, 98, 101, 106, 107, 108
    generic hybridity of 4, 11, 13, 84, 88, 94, 100
    male friendship in 83, 84, 85, 86, 88, 90, 91, 92, 93, 95, 97, 98, 100, 102, 105, 108
    significance of Roman historical setting of 12, 83, 84, 85, 86, 87, 89, 90, 91, 93, 94, 95, 96, 97, 98, 102, 105, 106
    tourism and tourists in 84, 85, 86, 87, 96, 102
"Andrea del Sarto" (Browning), 109
anthropology 2, 19*n*, 21*n*, 22, 23, 27, 142
"Apt Vogler" (Browning) 111
Appleton, Jay 80*n*
Aristophanes' Apology (Browning) 9
Armstrong, A. J. 113
Armstrong, Isobel 4*n*, 75*n*, 104, 113

Arnold, Matthew 2, 4, 7–9, 28, 48, 83, 88, 93, 103, 104
Arnold, Tom 100
Arnold, Thomas (father) 84
Auerbach, Nina 12*n*, 75*n*, 120, 146*n*
Aurora Leigh (Barrett Browning) 1, 4, 8, 10, 11, 12, 57–81, 88, 105, 108, 114, 120*n*, 126, 137, 142, 146; *see also* chivalry; feminist criticism; heroism
    as *Bildungsroman* and *Kunstlerroman* 58, 60, 75, 79
    fatherhood in 60, 62–5, 67, 68, 69, 70, 72, 74, 75, 76, 78, 81
    gender role reversal in 60, 64, 78, 79
    female-female competition in 60, 61, 68, 69, 70, 72, 73, 74, 75, 76, 77, 145
    male love in 62, 63, 64, 65, 66, 67, 68, 70, 72, 73, 74, 76, 77, 78, 79, 80
    religion and spirituality in 62, 63, 66, 67, 69, 73, 74, 79, 80
    Romney Leigh's masculinity in 57, 58, 60, 62, 66, 67, 69, 70, 71, 72, 74, 78
    as verse-novel 58, 62, 78, 81
Austen, Jane 20*n*, 24, 107
Avery, Simon 58
Aytoun, William Edmonstoune 5, 8, 78

Bailey, P. J. 4, 5
Balaustion's Adventure (Browning) 9
Barrett, Edward Moulton- (Barret Browning's father) 59, 62
Barrett, Edward Moulton- (Barret Browning's brother) 58, 72
Barrett Browning, Elizabeth 1, 2, 4, 5, 8, 9, 10, 11, 12, 21, 26, 28, 57–81, 82, 84, 88, 106, 107, 108, 109, 110, 111, 112, 114, 115, 118, 126, 133*n*, 135, 140, 141, 143, 145, 146; *see also specific works by*

critical reputation as poet 10, 11, 12, 57, 58, 59, 65n, 70–71, 76, 78–81
interest in epic genre and traditions 58, 59, 60, 73, 77, 78, 79, 81
marriage to Browning 59, 62, 71n
relationship with father 59, 62
Spasmodic poetry and 12, 67, 78, 79, 81
The Battle of Marathon (Barrett Browning) 9, 58
Beer, Gillian 2
Bells and Pomegranates (Browning), 109
biocultural studies 20, 22, 23, 25n, 28
biopoetics 22, 24
Biswas, Robindra Kumar 88n, 99n, 104, 105, 106
Blair, Kirstie 5n, 78
Boos, Florence S. 5n
Booth, Wayne C. 24
The Bothie of Tober-na-Vuolich (Clough) 9, 84, 88, 99, 100, 101, 102, 105
Boyd, Brian vi, 19–20, 21, 22, 23, 24, 28, 142n
Boyd, Hugh Stuart 71
Brady, Ann P. 120
Brisbane, Thomas, 6
Brooks, Peter 138
Brown, Donald E. 19, 27
Browning, Robert 1, 2, 4, 9, 10, 11, 12, 17, 21, 26, 28, 59, 62, 88, 104, 106, 107, 109–40, 141, 144, 145, 146, 147; *see also specific works by*
conflicted masculinity of 119, 120
courtship of and marriage to Barrett Browning 110, 111, 112, 118, 120
critical reputation as poet 109, 110, 111, 112, 113, 118, 119, 120, 124, 133
critique of conventional masculinity 109, 110, 119, 120, 121, 125, 127, 129, 131, 132, 133, 134, 135, 136, 137, 139, 140, 144, 145, 146, 147
idealization of women 110, 111, 118, 124, 133, 136, 137, 139, 142, 144, 145
Browning, Robert Wiedemann (Penini or Pen) 59, 110, 111
Buckler, William E. 119, 127, 136
Buckley, Jerome H. 13, 50
Burbidge, Thomas 100
Buss, David M. 19n, 143

"Caliban upon Setebos" (Browning) 111
Carroll, Joseph vi, 20, 21n, 22, 23, 24, 25, 26, 27, 28, 107.
Carlyle, Thomas 14, 15, 39, 42, 66, 94
Casa Guidi Windows (Barrett Browning) 9, 59
Case, Alison 62
Chaucer, Geoffrey 5, 103
Chesterton, G. K. 111
chivalry 29, 36–9, 42, 46–8, 51, 52, 55, 69, 74, 76, 78, 89, 90, 122, 125, 126, 128, 131–3, 137, 139, 142, 145, 146; *see also* heroism
Chorley, Katharine 104, 105
Christ, Carol 12
Christiansen, Rupert 102n
Christmas Eve and Easter Day (Browning) 9, 110
Clough, Arthur Hugh 1, 2, 4, 5, 7, 8, 9, 11, 12, 13, 17, 21, 26, 28, 83–108, 114, 139, 140, 141, 144, 145, 146; *see also specific works by*
critical reputation as poet 9, 83, 88, 98, 99, 100, 102, 103, 104, 105, 106
identification with protagonist in *Amours* 11, 83, 84, 85, 86, 87, 95, 105, 106, 107, 108
relationship with Matthew Arnold and the Arnold family 78, 83, 88, 93, 98, 100, 103, 104, 105
ties to Oxford and sense of vocation 83, 85, 95, 98, 100, 102, 108
Clough, Blanche Smith (Clough's wife) 102, 103, 104n, 105, 106
cognitive science 16, 22, 24, 25, 27
consilience, 3
Cook, A. K. 124n
Cooke, Brett vi, 22, 23, 24
Cooper, Helen 57n, 58
Corrigan, Beatrice 124n, 134n
Cott, Nancy F. 44
"Count Gismond" (Browning) 110, 126
cowardice 91, 97, 106, 107, 121, 122, 123, 124, 125, 126, 127, 128, 131, 132, 133, 136, 145, 146
Crowder, Ashby 119
"The Cry of the Children" (Barrett Browning) 59

Dalley, Lana 80

Darwin, Charles 2, 22, 132; *see also* evolutionary (Darwinian) theory; literary Darwinism
David, Deirdre 57*n*
"A Death in the Desert" (Browning) 111
Dentith, Simon 12, 13, 17, 37, 81
Derrida, Jacques 13
DeVane, William C. 112, 132, 133
Dipsychus (Clough) 9, 99, 102
Dissanayake, Ellen 23, 26
"Dover Beach" (Arnold) 48
Dramatic Lyrics (Browning) 109
Dramatis Personae (Browning) 111
Dutton, Denis 23, 29

Eagleton, Terry 50*n*
The Earthly Paradise (Morris) 5, 9
Easterlin, Nancy vi, 23, 24*n*, 25, 79
ecocriticism 25, 26, 79
Edwin of Deira (Smith) 8, 103
Eliot, George 2
Eliot, T. S. 12*n*, 31, 50, 51, 99, 112
Emerson, Ralph Waldo 93, 98, 102
Enoch Arden (Tennyson) 8
epic
    classical or traditional 4, 12, 17, 31, 37, 52*n*, 58, 84, 88, 89, 100, 114
    Victorian long poem as 3–13, 31, 37, 52, 58, 59, 60, 73, 77, 78, 79, 81, 84, 88, 89, 99, 100, 103, 109, 114, 124*n*, 147
    mock-epic 4, 13, 100
An Essay on Mind (Barrett Browning) 9
evolutionary psychology 18, 20, 21, 22, 24, 27, 28, 29, 142; *see also* adaptationist psychology; cognitive science
evolutionary theories of art 22, 23, 24, 26, 142; *see also* theory of mind
evolutionary (Darwinian) theory 2, 19, 20, 21, 27, 67

Farrell, John P. 9*n*
Fauconnier, Gilles 25
feminist criticism 1, 3, 13–15, 31, 42*n*, 57, 65, 70, 71, 75*n*, 81, 113, 119, 120, 127, 133, 138, 139
Festus (Bailey) 4, 5, 9
Fifine at the Fair (Browning) 9
Fisher, Maryanne 70*n*

Foucault, Michel 13, 14
Fox, Robin 23, 52*n*
"Fra Lippo Lippi" (Browning), 109
Fussell, Paul 51*n*

Gavins, Joanna 25
Geary, David G. 17*n*, 19*n*, 27, 28, 52*n*, 77
gender 1, 10, 12–19, 31, 34–7, 42, 45, 49, 57, 58, 60, 64, 65, 72, 73, 78–80, 106, 110, 111, 113, 114, 118–20, 124, 128, 133, 137, 138, 142; *see also* feminist criticism; masculinity; men's studies; sexuality
genetic determinism 2, 28
Gest, John Marshall 113*n*, 124*n*, 134*n*
Gilbert, Elliot L. 31, 34, 41, 44, 46
Gilbert, Sandra 13
Gilfillan, George 6
Girouard, Mark 36*n*, 37*n*, 39, 47*n*, 51*n*
Gottschall, Jonathan 18, 20, 23, 24*n*
Graham, Colin 12, 13
Greenberger, Evelyn Barish 104
Gubar, Susan 13

Hall, Donald E. 18
Hallam, Arthur Henry 8, 32, 33
Hanley, Evelyn 58, 65*n*
Hardy, Thomas 2
Harris, Wendell V. 104
Harrison, Antony, 5*n*, 14
heroism 12, 13, 18, 25, 31, 32, 33, 38, 39, 52, 53, 54, 55, 65, 69, 70, 71, 74, 76, 77, 78, 79, 80, 81, 90, 92, 96, 98, 99, 100, 107, 108, 112, 114, 122, 124, 126, 128, 132, 136, 143, 144, 145, 146; *see also* chivalry
Hogan, Patrick Colm 23*n*, 25, 26, 27, 33, 53, 55*n*, 93
Holloway, John 13
Homans, Margaret 13
Homer 2, 4, 17–18, 88, 141
Honan, Park 11*n*, 111*n*, 118
Houghton, Walter 13, 14, 15, 50, 90, 94, 104, 105
Hughes, Linda K. 1*n*, 5*n*
human nature 1, 2, 14, 16, 19–30, 37, 39, 46, 50, 51, 52, 80, 89, 94, 99, 106, 107, 114, 135, 140, 141, 142, 143, 144, 146, 147; *see also* universals (human)

Idylls of the King (Tennyson) 1, 4, 8, 10, 11, 12, 16, 29, 31–55, 90, 103, 114, 133, 141, 142, 146, 147; *see also* chivalry; heroism; human nature; male violence
   adultery of Guinevere and Lancelot in 34, 35, 39, 43, 45, 47, 49, 52, 54, 142, 143
   Arthur's manhood in 31, 32, 33, 34, 37, 38, 41, 47, 48, 52
   concepts of historical and mythic time in 12, 37, 40, 41, 42, 47, 49, 51, 52, 53, 55, 146
   knighthood and chivalric codes in 36, 37, 38, 39, 42, 43, 44, 45, 46, 47, 48, 50n, 51 52, 53, 54, 55
   madness and bestiality in 34, 38, 39, 40, 41, 43, 44, 45, 47
   marriage of Arthur and Guinevere in 33, 42, 43, 44, 45, 52, 53, 54
   spirituality and Christian idealism in 32, 33, 37, 38, 40, 41, 44, 46, 47, 49, 50, 52, 54, 55
   idealization of women in 36, 37, 39, 40, 42, 43, 44, 45, 46, 47
In Memoriam A. H. H. (Tennyson) 9, 16, 33
The Inn Album (Browning) 9
Irvine, William 11n, 111n, 118

James, Henry 31, 99, 111, 119
"James Lee's Wife" (Browning) 111
Jobling, Ian 23, 70n
Johnson, E. D. H. 50n
Johnson, John A. 20
Johnson, Mark 80n
Johnson, Stephanie L. 65n

Kaplan, Cora 13, 58, 77n
Kaplan, Hillard 21
Karlin, Daniel 137n, 139
Kenny, Anthony 83n, 85, 87n, 104, 105, 106
Kenyon, John 59, 113
Kingsley, Charles 15, 38, 71, 78, 80n
Knoepflmacher, U. C. 119
Kruger, Daniel J. 20, 70n

"Lady Geraldine's Courtship" (Barrett Browning) 59
"The Lady of Shalott" (Tennyson) 32, 33, 34

Lakoff, George 80n
Langbaum, Robert 112, 119, 132
LaPorte, Charles 5n, 7n
"The Latest Decalogue" (Clough) 104
Leavis, F. R. 112
Leighton, Angela 75n
Levine, George 13
"A Life-Drama" (Smith) 5–8, 78, 79
literary Darwinism 1, 2, 18, 19–29, 81, 107, 108; *see also* biocultural studies, biopoetics
Loucks, James F. 119, 133n
Love, Glen A. 25, 26
Lowry, Howard Foster 103

McEwan, Ian 141
McLaren, Angus 18
McNally, James 120n
Madden, William 13
male violence
   in *Amours de Voyage* 91, 107, 145
   in *Aurora Leigh* 76, 77, 145, 147
   in *Idylls of the King* 34–7, 39, 41–3, 48, 51–2, 54–5, 142, 145
   innate potential for 16–18, 142
   in *The Ring and the Book* 121–3, 128, 131, 135, 137, 139, 145
   and the Victorian criminal justice system, 137–9
Malory, Sir Thomas 31, 32, 38, 39, 48n, 52, 143
Mari Magno (Clough) 103
Markovits, Stephanie 99
masculinity (critical and theoretical assumptions about) 1, 3, 13–19, 29, 135, 142, 147;
   see also gender; male violence; men's studies, sexuality; *specific works and authors*
Mason, Emma 5n
Maud (Tennyson) 8, 41
Men and Women (Browning), 110
men's studies 14, 15
Meredith, George 7
Mermin, Dorothy 58, 70n, 74n, 75n
Miller, Betty 112, 119, 133n
Millet, Kate 35
"Mr. Sludge, the Medium" (Browning) 111
Mind, theory of 24, 27, 52, 77, 139
Morgan, Thaïs 15

Morris, William 5, 44
"Morte d'Arthur" (Tennyson) 10, 11, 33, 34, 53
Mulhauser, F. L. 103, 104, 149
Munich, Adrienne 132–3, 135, 136
"My Last Duchess" (Browning), 109, 118, 119, 129

"Natura Naturans" (Clough) 100–101
Nelson, Claudia 75n
Nesse, Margaret 23
Newton, Judith 13
Nicolson, Harold 51
Nordlund, Marcus 24n, 144n
Norrington, A. L. P. 103

O'Neill, P. 112n

Pade, W. D. 32n
Palmer, Alan 25
Paracelsus (Browning) 109
Parleyings with Certain People (Browning) 9
Pater, Walter 15, 42
Patmore, Coventry 42
Pauline (Browning), 109
Phillips, Catherine 50n
Pinker, Steven 17, 19n
Pippa Passes (Browning), 109
Poovey, Mary 13
"Porphyria's Lover" (Browning), 109
postmodernism 1, 25, 29, 113
poststructuralism 13
Pound, Ezra 12n, 112
Prince Hohenstiel-Schwangau (Browning) 9
The Princess (Tennyson) 9, 35

Rabkin, Eric S. 23, 24
Rampton, David 2n
Red Cotton Night-Cap Country (Browning) 9
Reynolds, Matthew 9, 141n
Ricks, Christopher 31
Ridenour, George M. 120n
The Ring and the Book (Browning) 1, 4, 9, 10, 11, 12, 109–40, 144, 145, 146, 147; *see also* chivalry; cowardice; heroism; male violence
  bachelorhood and male celibacy in 116, 117, 127, 136, 137, 139, 146, 147

  Christianity and the Church in 112, 113, 114, 115, 117, 120, 123, 124n, 125, 126, 128, 129, 130, 131, 132, 133, 134, 135, 137, 139, 142, 145, 146
  feminist interpretations of 113, 119, 120, 127, 133, 135, 136, 138, 139
  heroism and chivalry of Caponsacchi in 115, 120, 122, 124, 125, 126, 128, 131,132,136, 137, 146
  idealization of Pompilia in 111, 115, 118, 120, 124n, 126, 133n, 134, 136, 137
  flawed masculinity of Guido in 120, 121, 122, 123, 124, 125, 126, 127, 128, 129, 131, 132, 133, 134, 135, 136, 137, 139, 145
  the Old Yellow Book as source for 113, 121, 124n, 134, 135
  the Pope as Browning's spokesman in 128, 129
Roberts, Adam 9n
Robinson, Henry Crabb 31
Rosenberg, John D. 32n, 34, 40, 46n, 50, 51
Rudy, Jason R. 5n
Ruskin, John 37, 42, 57, 70
Ryals, Clyde de L., 45n, 46, 50, 111n, 118n

"Say Not the Struggle Nought Availeth" (Clough) 104–5
Scott, Patrick 88n, 95, 104, 149
Sedgwick, Eve 14
sexuality 14, 15, 17, 18, 19, 23, 28, 34, 35, 38, 42, 43, 44, 45, 46, 47, 48, 52, 68, 70, 71, 72, 73, 75, 76, 81, 86, 101, 102, 106, 107, 119, 120, 126, 127, 129, 140, 142, 143, 145
Shairp, John Campbell 98
Shaw, Marion 35
Shaw, W. David 4n, 31n, 113
Showalter, Elaine 13
Simon, Carl P. 23
Simpson, Roger 32
Slingerland, Edward 24n
Slinn, E. Warwick 4n, 113
Smith, Alexander 5–8, 26, 78, 79, 103
social construction 16, 29, 49, 52, 80, 143n, 144
"Soliloquy of the Spanish Cloister" (Browning), 109

Sonnets from the Portuguese (Barrett Browning) 59
Sordello (Browning), 109

Spasmodic poetry 5–8, 12, 26, 67, 78, 79, 81
Spencer, Herbert 7
Steen, Gerald 25
Stockwell, Peter 26
Stone, Marjorie 58
Storey, Robert 23, 25
Stott, Rebecca 58
Strand, Mark 9*n*
Sugiyama, Michelle Scalise 23, 24*n*
Sussman, Herbert 14, 15, 16, 18, 28, 31*n*, 35, 39, 42
Swedenborg, Emmanuel 63, 66, 74, 80
Swinburne, Algernon Charles 31, 38, 39, 52

Taylor, Beverly 14
Tennyson, Alfred, Lord 1, 2, 3*n*, 4, 5, 8, 9, 10, 11, 12, 16, 17, 21, 26, 28, 29, 31–55, 59, 60, 69, 71, 90, 103, 104, 107, 108, 114, 132, 133, 135, 140, 141, 142, 143, 145, 146; *see also specific works by*
   complex attitudes toward masculinity 31, 32, 33, 34, 35, 37, 38, 40, 41, 42, 46, 47, 48, 51, 52, 55, 71, 90, 132, 133, 135, 140, 142, 145, 146
   critical reputation as poet 8, 31, 32, 33, 34, 35, 37, 38, 43, 46, 49, 50, 51, 52, 55, 103, 114
   family relationships 31, 32, 33, 41
   interest in the Arthurian legend 1, 10, 11, 17, 31, 32, 33, 34, 37, 38, 39, 40, 41, 43, 46, 47, 48, 49, 51, 52, 141, 146
   portrayal of romantic love 33, 35, 39, 52*n*, 53
Tennyson, Hallam 1*n*, 31, 33, 47, 49
Thorpe, Michael 104
Timko, Michael 99, 104
Tosh, John 15, 38*n*, 75*n*
Trilling, Lionel 2*n*
Tucker, Herbert F. 5*n*, 12, 13, 79, 81, 113, 114, 124
Turner, Frederick 22, 23, 24
Turner, Mark 25

universals (human) 2, 12, 14, 18, 19, 20, 23, 25, 26, 27, 29, 30, 33, 37, 50, 52, 81, 89, 93, 99, 106, 107, 114, 133, 135, 141, 142, 146, 147; *see also* human nature

Vicinus, Martha 13
Victorian studies 1, 13–19, 57
Virgil, 89

Wade, Nicholas 21*n*
Weinstein, Mark A. 5*n*
Wiener, Martin J. 18, 137–9
Williams, Raymond 15
Wilson, David Sloan 23, 24, 52
Wilson, E. O. 3*n*, 26, 27
Woolford, John 137*n*, 139
Wörn, Alexandra M. B. 80
"The Worst of It" (Browning) 111

Zonana, Joyce 58
Zunshine, Lisa 24, 28